Areteem Institute

Math Challenge II-B Combinatorics

Edited by John Lensmire
David Reynoso
Kevin Wang
Kelly Ren

WWW.ARETEEM.ORG

PUBLISHED BY ARETEEM PRESS

ISBN: 1-944863-36-2
ISBN-13: 978-1-944863-36-4
First printing, November 2018.

TITLES PUBLISHED BY ARETEEM PRESS

Cracking the High School Math Competitions (and Solutions Manual) - Covering AMC 10 & 12, ARML, and ZIML
Mathematical Wisdom in Everyday Life (and Solutions Manual) - From Common Core to Math Competitions
Geometry Problem Solving for Middle School (and Solutions Manual) - From Common Core to Math Competitions
Fun Math Problem Solving For Elementary School (and Solutions Manual)

ZIML MATH COMPETITION BOOK SERIES

ZIML Math Competition Book Division E 2016-2017
ZIML Math Competition Book Division M 2016-2017
ZIML Math Competition Book Division H 2016-2017
ZIML Math Competition Book Jr Varsity 2016-2017
ZIML Math Competition Book Varsity Division 2016-2017
ZIML Math Competition Book Division E 2017-2018
ZIML Math Competition Book Division M 2017-2018
ZIML Math Competition Book Division H 2017-2018
ZIML Math Competition Book Jr Varsity 2017-2018
ZIML Math Competition Book Varsity Division 2017-2018

MATH CHALLENGE CURRICULUM TEXTBOOKS SERIES

Math Challenge I-A Pre-Algebra and Word Problems
Math Challenge I-B Pre-Algebra and Word Problems
Math Challenge I-C Algebra
Math Challenge II-A Algebra
Math Challenge II-B Algebra
Math Challenge III Algebra
Math Challenge I-A Geometry
Math Challenge I-B Geometry
Math Challenge I-C Topics in Algebra
Math Challenge II-A Geometry
Math Challenge II-B Geometry
Math Challenge III Geometry
Math Challenge I-A Counting and Probability
Math Challenge I-B Counting and Probability
Math Challenge I-C Geometry
Math Challenge II-A Combinatorics
Math Challenge II-B Combinatorics

Math Challenge III Combinatorics
Math Challenge I-B Number Theory
Math Challenge II-A Number Theory

COMING SOON FROM ARETEEM PRESS

Fun Math Problem Solving For Elementary School Vol. 2 (and Solutions Manual)
Counting & Probability for Middle School (and Solutions Manual) - From Common Core to Math Competitions
Number Theory Problem Solving for Middle School (and Solutions Manual) - From Common Core to Math Competitions
Other volumes in the **Math Challenge Curriculum Textbooks Series**

The books are available in paperback and eBook formats (including Kindle and other formats). To order the books, visit `https://areteem.org/bookstore`.

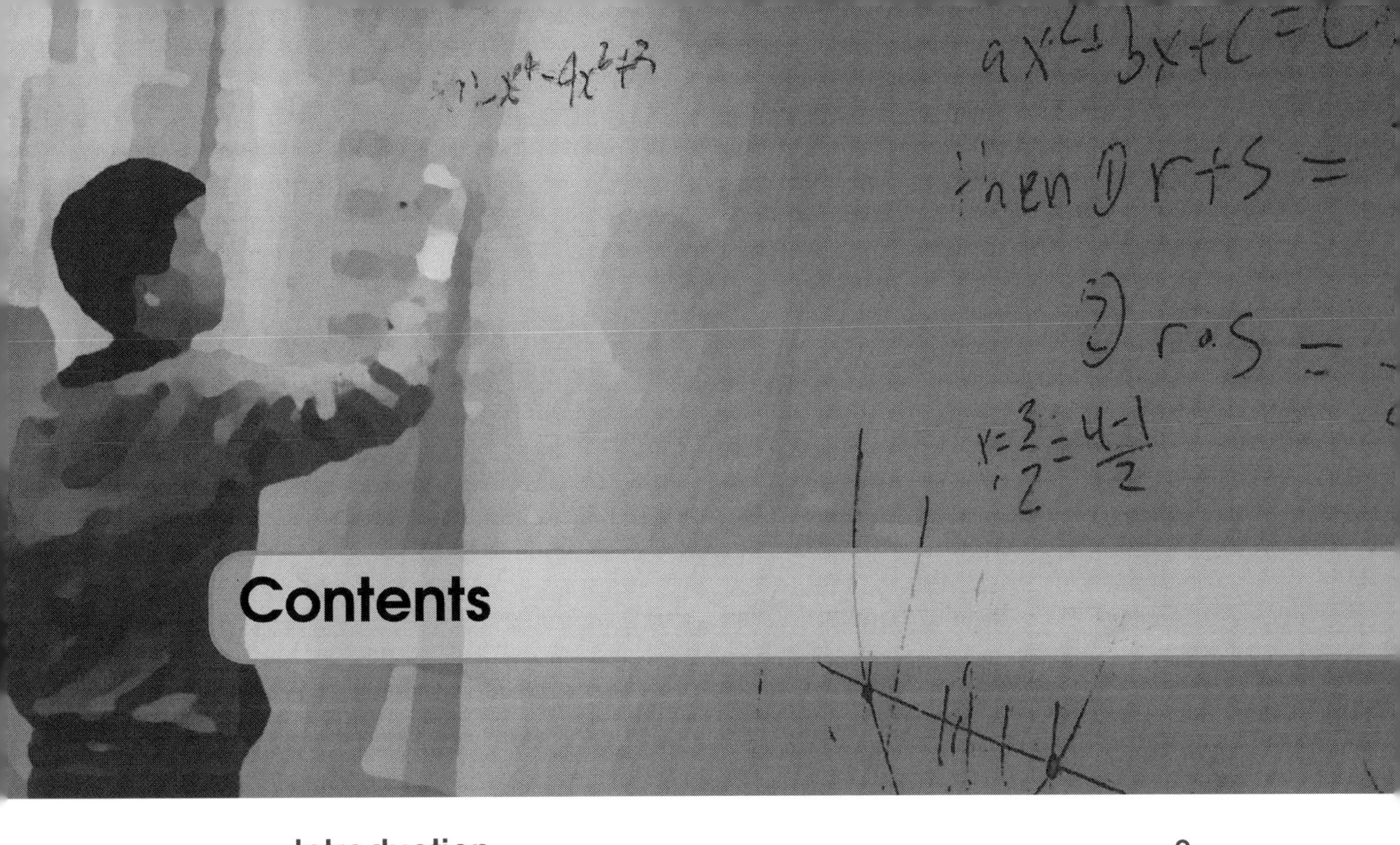

Contents

Introduction

The math challenge curriculum textbook series is designed to help students learn the fundamental mathematical concepts and practice their in-depth problem solving skills with selected exercise problems. Ideally, these textbooks are used together with Areteem Institute's corresponding courses, either taken as live classes or as self-paced classes. According to the experience levels of the students in mathematics, the following courses are offered:

- Fun Math Problem Solving for Elementary School (grades 3-5)
- Algebra Readiness (grade 5; preparing for middle school)
- Math Challenge I-A Series (grades 6-8; intro to problem solving)
- Math Challenge I-B Series (grades 6-8; intro to math contests e.g. AMC 8, ZIML Div M)
- Math Challenge I-C Series (grades 6-8; topics bridging middle and high schools)
- Math Challenge II-A Series (grades 9+ or younger students preparing for AMC 10)
- Math Challenge II-B Series (grades 9+ or younger students preparing for AMC 12)
- Math Challenge III Series (preparing for AIME, ZIML Varsity, or equivalent contests)
- Math Challenge IV Series (Math Olympiad level problem solving)

These courses are designed and developed by educational experts and industry professionals to bring real world applications into the STEM education. These programs are ideal for students who wish to win in Math Competitions (AMC, AIME, USAMO, IMO,

ARML, MathCounts, Math League, Math Olympiad, ZIML, etc.), Science Fairs (County Science Fairs, State Science Fairs, national programs like Intel Science and Engineering Fair, etc.) and Science Olympiad, or purely want to enrich their academic lives by taking more challenges and developing outstanding analytical, logical thinking and creative problem solving skills.

In Math Challenge II-B, students learn and practice in areas such as algebra and geometry at the high school level, as well as advanced number theory and combinatorics. Topics include polynomials, inequalities, special algebraic techniques, trigonometry, triangles and polygons, collinearity and concurrency, vectors and coordinates, numbers and divisibility, modular arithmetic, residue classes, advanced counting strategies, binomial coefficients, and various other topics and problem solving techniques involved in math contests such as the American Mathematics Competition (AMC) 10 & 12, ARML, beginning AIME, and Zoom International Math League (ZIML) Junior Varsity and Varsity Divisions.

The course is divided into four terms:

- Summer, covering Algebra
- Fall, covering Geometry
- Winter, covering Combinatorics
- Spring, covering Number Theory

The book contains course materials for Math Challenge II-B: Combinatorics.

We recommend that students take all four terms. Each of the individual terms is self-contained and does not depend on other terms, so they do not need to be taken in order, and students can take single terms if they want to focus on specific topics.

Students can sign up for the live or self-paced course at `classes.areteem.org`.

About Areteem Institute

Areteem Institute is an educational institution that develops and provides in-depth and advanced math and science programs for K-12 (Elementary School, Middle School, and High School) students and teachers. Areteem programs are accredited supplementary programs by the Western Association of Schools and Colleges (WASC). Students may attend the Areteem Institute in one or more of the following options:

- Live and real-time face-to-face online classes with audio, video, interactive online whiteboard, and text chatting capabilities;
- Self-paced classes by watching the recordings of the live classes;
- Short video courses for trending math, science, technology, engineering, English, and social studies topics;
- Summer Intensive Camps held on prestigious university campuses and Winter Boot Camps;
- Practice with selected free daily problems and monthly ZIML competitions at ziml.areteem.org.

Areteem courses are designed and developed by educational experts and industry professionals to bring real world applications into STEM education. The programs are ideal for students who wish to build their mathematical strength in order to excel academically and eventually win in Math Competitions (AMC, AIME, USAMO, IMO, ARML, MathCounts, Math Olympiad, ZIML, and other math leagues and tournaments, etc.), Science Fairs (County Science Fairs, State Science Fairs, national programs like Intel Science and Engineering Fair, etc.) and Science Olympiads, or for students who purely want to enrich their academic lives by taking more challenging courses and developing outstanding analytical, logical, and creative problem solving skills.

Since 2004 Areteem Institute has been teaching with methodology that is highly promoted by the new Common Core State Standards: stressing the conceptual level understanding of the math concepts, problem solving techniques, and solving problems with real world applications. With the guidance from experienced and passionate professors, students are motivated to explore concepts deeper by identifying an interesting problem, researching it, analyzing it, and using a critical thinking approach to come up with multiple solutions.

Thousands of math students who have been trained at Areteem have achieved top honors and earned top awards in major national and international math competitions, including Gold Medalists in the International Math Olympiad (IMO), top winners and qualifiers at the USA Math Olympiad (USAMO/JMO) and AIME, top winners at the

Zoom International Math League (ZIML), and top winners at the MathCounts National Competition. Many Areteem Alumni have graduated from high school and gone on to enter their dream colleges such as MIT, Cal Tech, Harvard, Stanford, Yale, Princeton, U Penn, Harvey Mudd College, UC Berkeley, or UCLA. Those who have graduated from colleges are now playing important roles in their fields of endeavor.

Further information about Areteem Institute, as well as updates and errata of this book, can be found online at `http://www.areteem.org`.

Acknowledgments

This book contains many years of collaborative work by the staff of Areteem Institute. This book could not have existed without their efforts. Huge thanks go to the Areteem staff for their contributions!

The examples and problems in this book were either created by the Areteem staff or adapted from various sources, including other books and online resources. Especially, some good problems from previous math competitions and contests such as AMC, AIME, ARML, MATHCOUNTS, and ZIML are chosen as examples to illustrate concepts or problem-solving techniques. The original resources are credited whenever possible. However, it is not practical to list all such resources. We extend our gratitude to the original authors of all these resources.

1. Counting Fundamentals

Sum and Product Rule

- **Sequential Counting Principle (Product Rule)**: Suppose that a procedure can be broken down into k successive tasks. If there are n_1 ways to do the first task, and n_2 ways to do the second task after the first task has been done, and so on, then there are $n_1 \times n_2 \times \cdots \times n_k$ ways to do the procedure.
- **Additive Counting Principle (Sum Rule)**: Suppose we have tasks $T_1, T_2, \ldots, T_k$ that can be done in $n_1, n_2, \ldots, n_k$ ways, respectively, and no two of these tasks can be done at the same time, then there are $n_1 + n_2 + \ldots + n_k$ ways to do one of these tasks.
- **Note**: Both of these rules are "reversible". For example, suppose a procedure can be broken down into 2 successive tasks. If there are n ways to do the entire procedure and m ways to do the 2nd task, then there are $\frac{n}{m}$ ways of doing the 1st task.

Permutations and Combinations

- **Permutations**: Permutation means arrangement of things *in a certain order*. The number of permutations of r elements taken out of a set of n elements (without repeating) is denoted ${}_nP_r$:

$$ {}_nP_r = n(n-1)(n-2)\cdots(n-r+1) = \frac{n!}{(n-r)!}. $$

- **Combinations**: Combination means selection of things where *order does not matter*. The number of combinations of r elements taken out of a set of n elements is denoted ${}_nC_r$ or $\binom{n}{r}$:

$${}_nC_r = \binom{n}{r} = \frac{n(n-1)(n-2)\cdots(n-r+1)}{r!} = \frac{n!}{r!(n-r)!}.$$

- **Relationship with Pascal's Triangle**: Here is Pascal's triangle, written both in the usual way, and written with its terms expressed as combinations.

$$\begin{array}{ccccccc} & & & \binom{0}{0} & & & \\ & & \binom{1}{0} & & \binom{1}{1} & & \\ & \binom{2}{0} & & \binom{2}{1} & & \binom{2}{2} & \\ \binom{3}{0} & & \binom{3}{1} & & \binom{3}{2} & & \binom{3}{3} \end{array} \qquad \begin{array}{ccccccc} & & & 1 & & & \\ & & 1 & & 1 & & \\ & 1 & & 2 & & 1 & \\ 1 & & 3 & & 3 & & 1 \end{array}$$

In other words, the entries in Pascal's triangle equal to the corresponding entry in the triangle of combination coefficients on the left. We'll explore this in more detail later on in this book.

Problem Solving Strategies for Counting

- Solve a few tiny problems of the same sort and see if you can find a pattern.
- Break the problem into several independent steps. (Product Rule.)
- Break the problem into parts. (Sum Rule.)
- Use standard permutation or combination formulas.
- Sometimes it is easier to count what you *don't* want.
- Sometimes problems that seem totally different are actually equivalent.

1.1 Example Questions

Problem 1.1 Suppose John has 2 hats, 4 shirts, 1 jacket, 3 pairs of pants, 5 pairs of shorts, and 4 pairs of shoes.

(a) Suppose John makes an outfit consisting of a shirt, a pair of pants, and a pair of shoes. How many different outfits does he have?

(b) Repeat (a) if John *can* wear shorts instead of pants.

(c) Now suppose John can wear shorts or pants as in (b), *but* if he wears shorts, he will also wear a hat and possibly a jacket.

Problem 1.2

(a) Find the number of ways to choose an ordered set of 3 numbers from $\{1,2,3,4\}$.

(b) Find the number of ways to choose 3 numbers from the set $\{1,2,3,4,5,6\}$

Problem 1.3 Suppose you have a group of 10 people. How many different photographs are there of everyone lined up if:

(a) all the people look different?

(b) 2 of the people are identical twins who have dressed identically?

(c) 2 of the people are a couple and must stand next to each other?

(d) 2 of the people are sworn enemies and cannot stand next to each other?

Problem 1.4 How many 2-digit numbers are even? Give at least 2 different proofs.

Problem 1.5 How many 10-digit numbers:

(a) do not have the digit 0?

(b) do not have the digit 5?

(c) contain the digit 5 exactly 5 times?

(d) contain the digit 5 at least once?

Problem 1.6 How many factors of 2^{20} are larger than $5,000$?

Problem 1.7 Suppose you write out the numbers $1-1000$: $1,2,3,4,\ldots,1000$.

(a) How many digits have you written in total?

(b) What is the sum of all the numbers written?

(c) What is the sum of all the digits written?

Problem 1.8 Suppose you have a 5-letter word.

(a) How many possible words are there?

(b) How many of the words do not have consecutive consonants or consecutive vowels (the vowels are a, e, i, o, u)? (That is, suppose the letters alternate between consonant and vowel.)

(c) How many words have exactly two vowels?

(d) How many words have a block of three consecutive consonants and two consecutive vowels.

Problem 1.9 Suppose you have a standard 8×8 chessboard.

(a) How many ways can you place 8 mutually non-attacking rooks on the chessboard? (Consider the rooks to be identical.)

(b) How many ways can you place 2 mutually non-attacking rooks on the chessboard?

(c) How many ways can you place 7 mutually non-attacking rooks on the chessboard? (Consider the rooks to be identical.)

Problem 1.10 Telephone numbers in the Land of Nosix have 7 digits, and the only digits available are $\{0,1,2,3,4,5,7,8\}$. No telephone number may begin in $0,1$, or 5. Find the number of telephone numbers possible that meet the following criteria:

(a) you may have repeated digits.

(b) you may not have repeated digits.

(c) you may have repeated digits, but the phone number must be even.

(d) you may not have repeated digits, and the phone number must be odd.

1.2 Quick Response Questions

Problem 1.11 Which of the following expressions is equal to $\binom{n}{k}$?

(A) $\binom{k}{n}$
(B) $\binom{n}{n-k}$
(C) $\binom{n+k}{n-k}$
(D) None of the above

Problem 1.12 Which of the following expressions is equal to $\binom{n}{2}$?

(A) n
(B) 1
(C) $\frac{n(n-1)}{2}$
(D) $\frac{n!}{(n-2)!}$

Problem 1.13 Suppose 7 people run a race. If there are no ties, how many different outcomes are there?

(A) 7
(B) 7!
(C) $\binom{7}{7}$
(D) $\binom{7}{2}$

Problem 1.14 Suppose 8 different people compete in 4 different events where there is only one winner. How many different ways can all the events be won?

(A) 8^4
(B) $8 \cdot 4$
(C) $\frac{8!}{4!}$
(D) $\binom{8}{4}$

Problem 1.15 How many ways can 6 people each choose a piece of cake from 10 total (no sharing).

(A) 10^6
(B) $10 \cdot 6$
(C) $\frac{10!}{4!}$
(D) $\binom{10}{6}$

Problem 1.16 How many ways can a committee of 4 people be chosen from 7 total candidates?

(A) 7^4
(B) $7 \cdot 4$
(C) $\frac{7!}{3!}$
(D) $\binom{7}{4}$

Problem 1.17 How many five letter 'words' start and end with a vowel?

(A) $5 \cdot 2 + 26 \cdot 3$
(B) $5^2 \cdot 26^3$
(C) 26^5
(D) $\binom{26}{3} \cdot \binom{5}{2}$

Problem 1.18 How many eight letter 'words' have an equal number of consonants and vowels?

(A) $21^4 \cdot 5^4$
(B) $\binom{8}{4} \cdot 21^4 \cdot 5^4$
(C) $\binom{26}{5} \cdot \binom{5}{4}$
(D) $21 \cdot 4 + 5 \cdot 4$

Problem 1.19 How many 9 letter words have 4 consonants and all the vowels?

(A) $\binom{26}{4} \cdot \binom{5}{5}$
(B) $\frac{21!}{17!} \cdot 5!$
(C) $\binom{9}{4} \cdot 21^4 \cdot 5!$
(D) $21^4 \cdot 5^5$

Problem 1.20 Suppose a man has 3 sons and 2 daughters, and has 2 boys' schools and 3 girls' schools available to choose from. How many different ways can he send his children to school if his daughters do not want to attend the same school?

(A) $2^3 \cdot 3^2$
(B) $2^3 + 3^2$
(C) $\binom{3}{2}^2$
(D) $2^3 \cdot 3 \cdot 2$

1.3 Practice Questions

Problem 1.21 John has 2 hats, 4 shirts, 1 jacket, 3 pairs of pants, 5 pairs of shorts, and 4 pairs of shoes. Suppose an outfit always contains a shirt, legwear (either shorts or pants), and shoes. Further, an outfit may contain a hat and/or a jacket. How many outfits in total does John have?

Problem 1.22 Suppose 8 people say $A, B, C, \ldots, H$ run a race.

(a) How many different outcomes are there for the race? (Suppose there are no ties!)

(b) How many different ways are there to give out Gold, Silver, and Bronze Medals?

(c) Suppose the top 4 finishers in the race advance to the next race. How many different groups can advance?

Problem 1.23 Suppose you have a group of 8 people. How many different photographs are there of everyone lined up if:

(a) 3 of the people are identical triplets who have dressed identically?

(b) 3 of the people are a family, 2 parents and a child, and the family must stand together, with the child in between the parents?

Problem 1.24 How many 2-digit numbers are a multiple of 9? Give at least 2 proofs.

Problem 1.25 How many 10-digit numbers, made up of only even digits:

(a) are even?

(b) contain the digit 0 exactly 2 times?

Problem 1.26 How many factors of $2^{10} \cdot 3$ are larger than 750?

Problem 1.27 Suppose that Billy is reading a book. If you wrote out all the page numbers that Billy has read $(1, 2, 3, \ldots)$ you would end up writing 348 total digits. What page number is Billy on?

Problem 1.28 Suppose you have a 7-letter word where repetitions are allowed, but not consecutive repetitions. How many words do not start with two consecutive consonants?

Problem 1.29 How many ways can you place 4 mutually non-attacking rooks on the chessboard, using only the outer edges of the board? (Assume the rooks are identical.)

Problem 1.30 Telephone numbers in the Land of Nosix have 7 digits, and the only digits available are $\{0, 1, 2, 3, 4, 5, 7, 8\}$. No telephone number may begin in 0, 1, or 5. Find the number of telephone numbers possible that meet the following criteria:

(a) You may have repeated digits, but the phone number must contain at least ≥ 2 different digits.

(b) You may have repeated digits, but the last 4 digits must all be different. Note: the phone number $222-2345$ *is* allowed.

2. Counting Methods I

Review of Permutations and Combinations

- **Permutations**: Permutation means arrangement of things *in a certain order*. The number of permutations of r elements taken out of a set of n elements (without repeating) is denoted ${}_nP_r$:

$$ {}_nP_r = n(n-1)(n-2)\cdots(n-r+1) = \frac{n!}{(n-r)!}. $$

- **Combinations**: Combination means selection of things where *order does not matter*. The number of combinations of r elements taken out of a set of n elements is denoted ${}_nC_r$ or $\binom{n}{r}$:

$$ {}_nC_r = \binom{n}{r} = \frac{n(n-1)(n-2)\cdots(n-r+1)}{r!} = \frac{n!}{r!(n-r)!}. $$

Grouping and Spacing

- **Grouping**: If some objects must be together, think of them as a single object/group when arranging. Then arrange among those objects if necessary.
- For example, if you are taking a photo with a couple that wants to stand together, first think of the couple as one 'person' and arrange. Then multiply by 2! to account for the ways the couple can be reordered.
- **Spacing**: If some objects must be separated, arrange the other objects first, and then place the objects in the spaces created by the others.

- For example, if you are taking a photo with a group of two enemies that must stand apart, first arrange the other people, setting aside the two enemies. The other people create spaces for the enemies, and then place the enemies in those spaces.

2.1 Example Questions

Problem 2.1 There are five Golden retrievers, six Irish setters, and eight Poodles at the pound.

(a) How many total ways are there to choose 2 dogs (in no particular order)?

(b) How many ways can two dogs be chosen if they are not the same kind?

(c) How many ways can two dogs be chosen so that not both of them are Poodles.

Problem 2.2 Suppose you have a student group with 15 males and 10 females.

(a) How many ways are there to pick a group of 5 males and 5 females?

(b) How many ways are there to pick an Executive Committee of 5 members and a Party Planning Committee of 5 members? Members can be on both committees at once, but each committee must have at least one male and at least one female.

(c) Suppose you still need to pick an Executive Committee and Party Planning Committee (each with 5 members). This time, only the Executive Committee is required to have a member of each gender, but now members are *not* allowed to be on both committees at once.

Problem 2.3 Suppose you have a group of 8 people. How many different photographs are there of everyone lined up if:

(a) 2 of the people are identical twins and 3 of the people are identical triplets (the twins and triplets dress identically)?

(b) the 8 people are 3 singles, a couple and the last 3 of the people are a family (2 parents and a child). The couple must be together and the family must stand together, with the child in between the parents?

Problem 2.4 5 boys and 3 girls run in a race.

(a) If all the boys finish as a group, how many outcomes are there?

(b) If none of the girls finish right after another (2nd and 3rd, 5th and 6th, etc.), how many outcomes are there?

Problem 2.5 The following illustration is a map of a city, and you would like to travel from the southwest (lower left) to the northwest (upper right) of the city along the roads in the shortest possible distance.

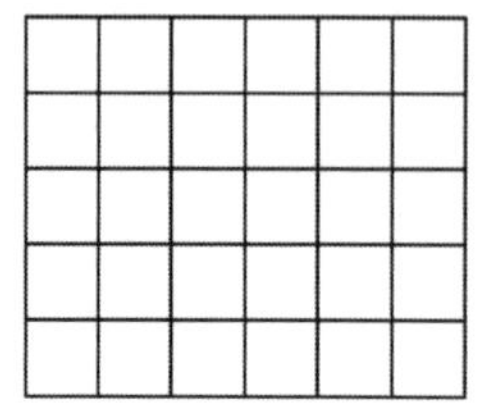

(a) In how many ways can you do this?

(b) Suppose you want to avoid the roads furthest to the west and north. How many different paths remain?

(c) How many rectangles are in the diagram?

Problem 2.6 Suppose in a group of soldiers you have 3 officers, 6 sergeants, and 30 privates.

(a) How many ways can a team be formed consisting of 1 officer, 4 sergeants, and 10 privates?

(b) Repeat (a) if the 6 sergeants each lead 5 different privates (with no overlap so the total number of privates is still 30) and privates can only be chosen if their lead sergeant is on the team as well.

Problem 2.7 Suppose you have 15 distinct playing cards. You want to divide them into 3 (unordered) groups. How many ways to do this are there if:

(a) the groups are of size 6, 5, and 4?

(b) all three groups have size 5?

Problem 2.8 A bookshelf contains 4 German books, 6 Spanish books, and 7 French books. Each book is different from one another.

(a) How many different arrangements can be done of these books?

(b) How many different arrangements can be done if books of each language must be next to each other?

(c) How many different arrangements can be done if no two German books must be next to each other?

Problem 2.9 Suppose a school offers the following classes: English, Latin, Algebra, Geometry, Calculus, History, Art, and Music. A student must take a total of 5 classes.

(a) If the order that the classes are chosen doesn't matter and at least one course must be a math course, how many different choices for classes does the student have?

(b) Suppose now the order of the classes does matter and that the student must take exactly one math class. How many choices for a schedule does the student have?

Problem 2.10 How many ways are there to arrange the numbers 21, 31, 41, 51, 61 so that the sum of each consecutive group of 3 numbers is divisible by 3?

2.2 Quick Response Questions

Problem 2.11 How many words are there of length 7 using the English alphabet?

(A) 26^7
(B) $\binom{26}{7}$
(C) $\frac{26!}{19!}$
(D) $26^7 - 21^2 \cdot 26^5$

Problem 2.12 How many words of length 6 alternate consonant-vowel-etc. or vowel-consonant-etc.?

(A) $\binom{21}{3} \cdot \binom{5}{3}$
(B) $2 \cdot \binom{26}{6}$
(C) $21 \cdot 5 \cdot 21 \cdot 5 \cdot 21 \cdot 5 + 5 \cdot 21 \cdot 5 \cdot 21 \cdot 5 \cdot 21$
(D) $21^3 \cdot 5^3$

Problem 2.13 How many ways are there to choose an unordered set of 6 different consonants and 3 different vowels from the English alphabet?

(A) $\binom{21}{6} \cdot \binom{5}{3}$
(B) $\binom{21}{6} + \binom{5}{3}$
(C) $21 \cdot 20 \cdot 19 \cdot 18 \cdot 17 \cdot 16 \cdot 5 \cdot 4 \cdot 3$
(D) $21^6 \cdot 5^3$

Problem 2.14 How many 7 letter words do not contain repeated letters?

(A) 26^7
(B) $\binom{26}{7}$
(C) $\frac{26!}{19!}$
(D) $26^7 - 21^2 \cdot 26^5$

Problem 2.15 How many 7 letter words do not start with two consecutive consonants?

(A) 26^7
(B) $\binom{26}{7}$
(C) $\frac{26!}{19!}$
(D) $26^7 - 21^2 \cdot 26^5$

Problem 2.16 How many rearrangements of the word *MACHINES* have no vowels next to each other?

Problem 2.17 How many words are there that contain 6 A's and 5 B's?

Problem 2.18 Suppose you have 4 males and 4 females. How many ways are there to line them all up so that the males are all together and the females are all together?

Problem 2.19 How many ways are there to pick a home team of 5 players and an away team of 5 players (no overlaps) from 20 total people?

(A) $\binom{20}{5} \cdot \binom{20}{5}$
(B) $\binom{20}{5} + \binom{20}{5}$
(C) $\binom{20}{5} \cdot \binom{15}{5}$
(D) $\binom{20}{5} + \binom{15}{5}$

Problem 2.20 How many ways are there to line up 10 people if three of the people are enemies and none of them can be next to each other?

(A) $10! - 3!$
(B) $\binom{10}{3} \cdot 10!$
(C) $7! \cdot 8 \cdot 7 \cdot 6$
(D) $7! \cdot 3!$

2.3 Practice Questions

Problem 2.21 There are now 7 Retrievers, 8 Setters, and 4 Poodles at the pound. Two of the dogs will be given a bath in succession.

(a) How many ways can this be done?

(b) Suppose that because of a Retriever's hair, it is impossible to wash a Setter or a Poodle if a Retriever is washed first. How many ways can two dogs be bathed now?

Problem 2.22 Suppose you have a student group with 10 seniors, 10 juniors, and 10 sophomores. You need to pick two different committees, each consisting of 6 students. Assume that a student can be on at most one committee. Each committee must have 3 seniors, 2 juniors, and 1 sophomore. How many ways are there to pick the committees?

Problem 2.23 8 people consisting of 4 couples line up for a photograph, with each couple together. How many photos are there with all the couples together?

Problem 2.24 6 boys and 4 girls run a race. Suppose the boys finish in alphabetical order based on their names. How many different outcomes of the race are there?

Problem 2.25 Consider the diagram below

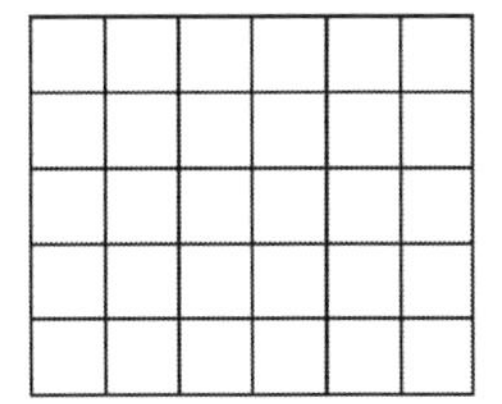

How many of the rectangles in the diagram are NOT squares?

Problem 2.26 Suppose you have a group of 3 officers, who each command 3 sergeants (no overlap). Further, each sergeant commands (no overlap) a set of 20 soldiers. (Therefore, each officer is indirectly in command of 60 soldiers.) A team is formed, consisting of 40 soldiers and a command unit. The command unit is either just 2 officers or just 6 sergeants. The 40 soldiers are chosen so that every soldier is commanded (directly or indirectly) someone in the command unit. How many different teams are possible?

Problem 2.27 Suppose 20 people will divide and form 4 teams of 5 people to play two games of 5 on 5 basketball, a first game and a second game. If we only care about the two games being played (i.e. Team A vs Team B and Team C vs Team D is the same as Team B vs Team A and Team D vs Team C) how many different ways can the games happen?

Problem 2.28 Suppose you have 6 German, 6 Spanish, and 5 French books on a bookshelf. How many ways are there to arrange the books if all the French books must be together, all the Spanish books must be together, but the French books and Spanish books cannot be next to each other?

Problem 2.29 Suppose a school offers 12 different classes: 3 in the History Department, 4 in the Math Department, 3 in the Literature Department, and 2 in the Arts Department. A day schedule includes 6 classes. If a student takes all 4 math classes, how many possible schedules are there?

Problem 2.30 How many ways are there to arrange the numbers $11, 21, 31, 41, 51, 61$ so that the sum each consecutive group of 3 numbers is divisible by 3?

3. Counting Methods II

Circular Permutations

- Suppose you have n distinct objects, arrange them on a circle. There are $\frac{n!}{n} = (n-1)!$ ways to do this. This is an example of a *circular permutation*.
- Note: In a circular permutation we only care about the arrangement of the objects. That is, the circular is allowed to rotate.
- For a general circular permutation, if we have n distinct objects and arrange r of them on a circle, there are $\left(\frac{n!}{(n-r)!}\right)/r = \frac{n!}{r\cdot(n-r)!}$ ways of doing so.

Stars and Bars

- Stars and Bars (Balls and Urns, etc., there are many different names) is a counting technique that should be memorized as all costs.
- **Non-Negative Version**: Given a positive integer n, and positive integer k, the number of ways to express n as the sum of k non-negative integers ($n = a_1 + a_2 + \cdots + a_k$ where $a_1, a_2, \ldots, a_k$ are non-negative) is $\binom{n+k-1}{n}$.
- **Positive Version**: Given a positive integer n, and positive integer k, the number of ways to express n as the sum of k positive integers ($n = a_1 + a_2 + \cdots + a_k$ where $a_1, a_2, \ldots, a_k$ are positive) is $\binom{n-1}{k-1}$.

3.1 Example Questions

Problem 3.1 Suppose 8 dinner guests attend a dinner party and are seated at a circular table. How many ways are there to seat the guests if:

(a) none of the seats are special, we only care about how they are arranged among themselves?

(b) one of the seats is the "Head of the Table"?

(c) none of the seats are special, but two of the guests are a couple and want to sit next to each other?

Problem 3.2 The number 3 can be expressed as a sum of one or more positive integers in four ways, namely, as 3, $1+2$, $2+1$, and $1+1+1$.

(a) How many ways can the number 5 be expressed as the sum of one or more positive integers less than or equal to 2?

(b) How many total ways can 5 be expressed as the sum of one or more positive integers?

(c) How many total ways can 5 be expressed as the sum of two non-negative integers?

Problem 3.3 Consider the number 6.

(a) How many ways can you express 6 as the sum of three non-negative integers?

(b) How many ways can you express 6 as the sum of three positive integers?

Problem 3.4 Beagel likes bagels, and he went to the Bagel Shop to buy 6 bagels for breakfast. The Bagel Shop sells 3 types of bagels: sourdough, blueberry, and sesame seeds.

(a) If Beagel plans to buy at least one of each type. In how many ways can he do this?

(b) If Beagel does not need to buy at least one of each type, how many ways can he buy the 6 bagels?

(c) How many ways are there to buy the bagels so that Beagel gets at least 2 types?

Problem 3.5 Suppose you have 30 identical balls and 6 numbered boxes. How many ways are there to put the balls into the boxes if:

(a) there are no restrictions?

(b) each box has at least two balls?

(c) no box has more than 5 balls?

(d) the first box has exactly 10 balls?

Problem 3.6 Consider the number 10000.

(a) How many factors does it have?

(b) How many ways are there to represent it as the product of 2 factors if we consider products that differ in the order of factors to be different?

(c) How many ways are there to represent it as the product of 3 factors if we consider products that differ in the order of factors to be different?

Problem 3.7 Suppose you have 5 blue, 5 red, and 5 green balls. You want to arrange the balls so that no two green balls are next to each other. How many ways are there to do this if

(a) the balls are in a row and each ball is numbered?

(b) the balls are in a row and each ball is identical?

(c) the balls are in a circle and each ball is numbered?

Problem 3.8 Suppose you have 30 numbered balls and 6 numbered boxes. How many ways are there to put the balls into the boxes if:

(a) there are no restrictions?

(b) no box has more than 5 balls?

Problem 3.9 Suppose 5 girls and 15 boys sit around the table. How many arrangements are there

(a) in total?

(b) if there is at least 1 boy in between all the girls?

Problem 3.10 Suppose 10 people get in an elevator on Floor 0. The people leave the elevator somewhere between (inclusive) Floors 1 and Floor 5.

(a) If we only care about how many people get of at each floor, how many ways can the people get off?

(b) If we only care about what collection of floors the elevator stops on, how many different collections are there?

3.2 Quick Response Questions

Problem 3.11 Suppose we want to write the letters A, B, C, D along the outside of a Frisbee (a circular disk). How many ways are there to do so? List all the outcomes.

Problem 3.12 How many solutions to the equation $a+b+c+d=10$ are there if a, b, c, d are all non-negative integers?

Problem 3.13 How many ways can the number 7 be written as the sum of one or more positive integers?

Problem 3.14 How many ways can the number 6 be written as the sum of 3 or more positive integers?

Problem 3.15 Suppose you have 5 couples. How many ways are there to arrange them in a line if all the couples stand next to each other?

Problem 3.16 Suppose you have 5 couples (male, female). How many ways are there to arrange all couples in a line with each couple together if the final line must have no two males and no two females next to each other?

Problem 3.17 How many ways are there to put 10 identical balls into 4 numbered boxes if each box must have at least 1 ball?

Problem 3.18 How many ways are there to put 10 numbered balls into 4 numbered boxes (a box is allowed to be empty)?

Problem 3.19 How many ways are there to put 4 numbered balls into 10 numbered boxes if at most one ball can be in each box?

Problem 3.20 How many factors does the number 196 have?

3.3 Practice Questions

Problem 3.21 Suppose 8 dinner guests attend a dinner party and are seated at a circular table. Two of the guests are a couple that must sit together. If one of the seats is the Head of the Table, how many ways are there to seat the guests?

Problem 3.22 Consider the expression $1+3+9+27+81+243+729$ (which equals 1093). Suppose you can erase some, all, or none of the numbers. How many possible sums are possible? (If you erase all the numbers, say the sum is 0.)

Problem 3.23 How many solutions to the equation $a+b+c=10$ are there with $a,b,c \geq 2$?

Problem 3.24 Suppose you want to bring a collection of 12 sodas to a party. You can choose from 6 types (and all that matters is how many of each soda you bring). How many ways can you do this if

(a) there are no restrictions on the types of soda you bring?

(b) one of your friends at the party really likes Fanta (one of the 6 types) so you want to bring at least 6 Fanta's to the party?

Problem 3.25 Suppose you have 12 identical balls and 4 numbered boxes. How many ways are there to put the balls in the boxes if the first box has at least 3 balls, but no more than 5 balls?

Problem 3.26 How many different ways are there to represent 810000 as the product of 3 factors if we consider products that differ in the order of factors to be different?

Problem 3.27 In how many ways can a necklace be made using 5 identical red beads and 2 identical blue beads?

Problem 3.28 Suppose you have 30 numbered balls and 6 numbered boxes. How many ways are there to put the balls into the boxes if the first box has exactly 10 balls?

Problem 3.29 Suppose 5 girls and 15 boys sit around the table. How many arrangements are there if there is at least 2 boys in between all the girls?

Problem 3.30 Suppose 5 people get in an elevator on Floor 0. The people leave the elevator somewhere between (inclusive) Floors 1 and Floor 10.

(a) If we only care about how many people get of at each floor, how many ways can the people get off?

(b) Suppose the 5 people all get off at different floors. If we only care about what collection of floors the elevator stops on, how many different collections are there?

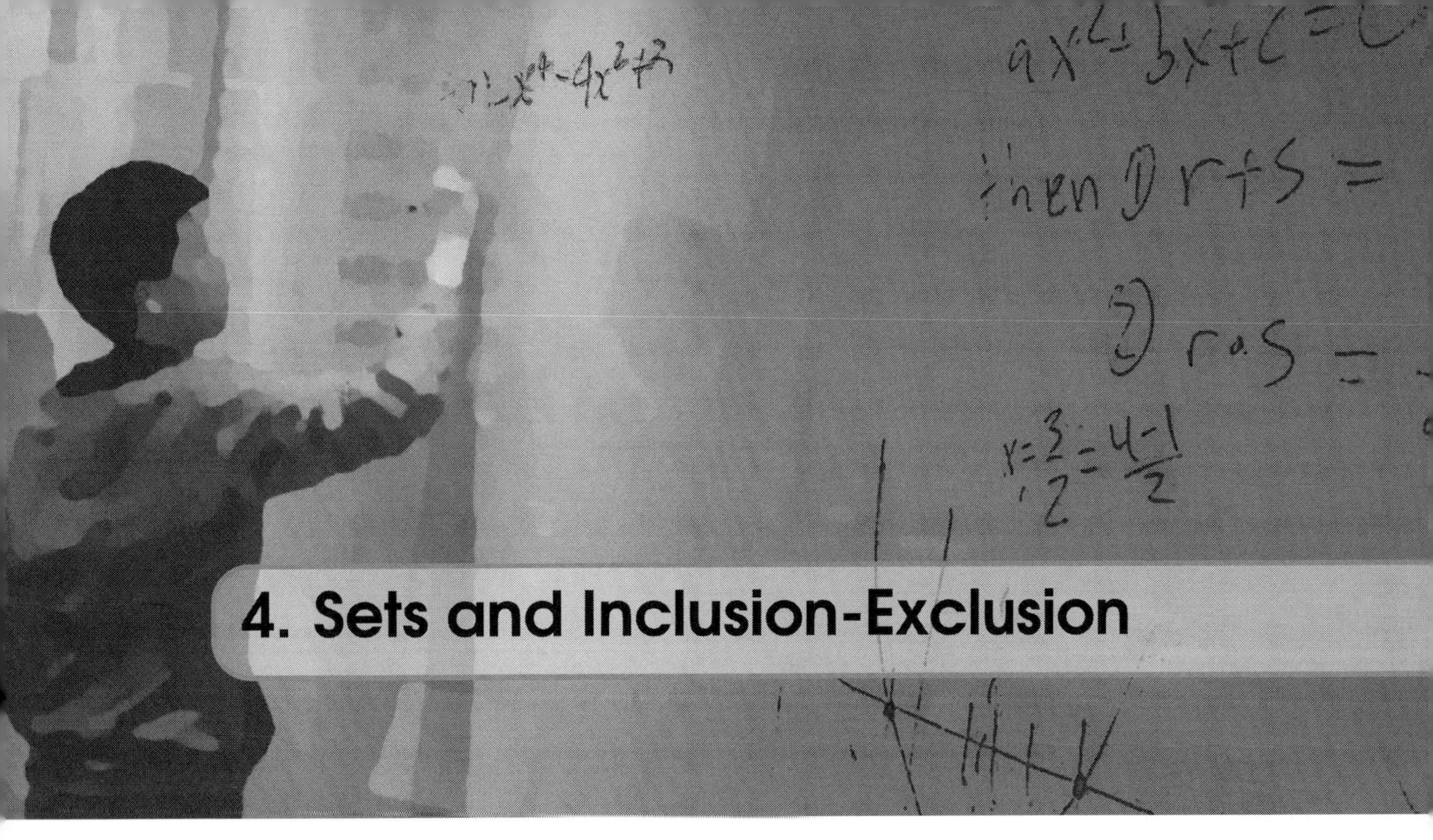

4. Sets and Inclusion-Exclusion

Basic Review of Sets

- A *set* is an unordered collection of elements, without repetitions. Sets are often denoted A, B, C, etc.
- If A, B are sets, then
  - $n(A), n(B)$ denotes the size of A and B respectively.
  - $A \cap B$ is the *intersection* of A and B. $A \cap B$ consists of all the elements both in A and in B.
  - $A \cup B$ is the *union* of A and B. $A \cup B$ consists of all the elements in A or in B (or in both).

Principle of Inclusion-Exclusion

- The *Principle of Inclusion-Exclusion* (or PIE for short) helps calculate the size of the union of two or more sets.
- For two sets, we have $n(A \cup B) = n(A) + n(B) - n(A \cap B)$.
- Venn Diagrams are useful in remembering and visualizing the Principle of Inclusion-Exclusion.

4.1 Example Questions

Problem 4.1 Suppose you have 8 pairs of people (so 16 people in total). How many ways are there to arrange the pairs (each pair next to each other) in a circle

(a) in total?

(b) if each pair is a different set of identical twins (that dress alike)?

(c) Repeat (a) if there is a special "Top" position in the circle that is different from the other places.

(d) Repeat (b) if there is a special "Top" position in the circle that is different from the other places.

Problem 4.2

(a) Work out and write out the PIE formula for 3 sets A,B,C.

(b) How many terms will the PIE formula for 4 sets A,B,C,D have?

Problem 4.3 Using the PIE formula, find the number of positive integers between 1 and 1000 that are either a multiple of 5, a multiple of 6, or a multiple of 7.

Problem 4.4 Suppose Albert, Bill, and Charles run a race.

(a) How many different outcomes of the race are there?

(b) Suppose now they run a second race. How many different outcomes for the 2nd race are there if no one finished the second race in the same place as the first race?

Problem 4.5 How many ways are there to put numbered 6 balls in 3 numbered boxes, so that each box gets at least one ball? Hint: It is probably easiest to use PIE and Complementary Counting.

Problem 4.6 Suppose you have a set $S = \{1,2,3,\ldots,40\}$. You want to choose A,B,C such that $A \cup B \cup C = S$ and $A \cap B \cap C = \emptyset$. (Remember $\emptyset = \{\}$ is the empty set, which contains no elements.) How many ways an this be done

(a) if we also assume $A \cap B = A \cap C = B \cap C = \emptyset$? (Under these conditions, A,B,C *partitions* S.)

(b) in total?

Problem 4.7 Suppose you have the numbers $\{1,2,3,4,5,6\}$.

(a) How many 6-digit numbers can be formed using each number once?

(b) How many 6-digit numbers can be formed with 2 next to 1 or 3? (Again use each number above once.) Hint: PIE.

(c) How many 6-digit numbers can be formed with 2 not next to 1 or 3? (Again use each number above once.)

Problem 4.8 Suppose that students take three tests in a course and that exactly 11 students get A's on each exam. How many students must get A's on all three exams if exactly 9 students get A's on any two exams and 14 students get an A on at least one exam?

Problem 4.9 Suppose some friends go to a party. They each wear a coat. However, as they are leaving, they each randomly grab a coat. How many ways can the friends leave so that *none* of them have their own coat, if there are:

(a) 3 friends?

(b) 4 friends?

Problem 4.10 Suppose you have copies of the 7 Harry Potter books. You give out the books to 3 of your friends. Each friend gets at least one book. How many ways can you give out the books? (The books you give each friend are not in any specific order.)

4.2 Quick Response Questions

Problem 4.11 Let $A = \{1,2,3,4,5\}$, $B = \{4,5,6,7,8\}$. What is $A \cap B$?

(A) $\{1,2,3,4,5,6,7,8\}$
(B) $\{1,2,7,8\}$
(C) $\{4,5\}$
(D) $\{4,5,6,7,8\}$

Problem 4.12 Let $A = \{1,2,3,4,5\}$, $B = \{4,5,6,7,8\}$. What is $A \cup B$?

(A) $\{1,2,3,4,5,6,7,8\}$
(B) $\{1,2,7,8\}$
(C) $\{4,5\}$
(D) $\{4,5,6,7,8\}$

Problem 4.13 Let $A = \{1,2,3,4,5\}$, $B = \{4,5,6,7,8\}$. What is $(A \cup B) \cap B$?

(A) $\{1,2,3,4,5,6,7,8\}$
(B) $\{1,2,7,8\}$
(C) $\{4,5\}$
(D) $\{4,5,6,7,8\}$

Problem 4.14 The Venn Diagram for two sets A, B has 3 sections. How many sections does a Venn Diagram for three sets A, B, C have?

Problem 4.15 How many numbers between 1 and 100 are multiples of either 2, 3, or 5?

Problem 4.16 How many numbers between 1 and 100 are a multiple of $4, 5$, or 6?

Problem 4.17 Suppose we want to write the letters A, B, C, D, E along the outside of a Frisbee (a circular disk). How many ways are there to do so? List all the outcomes.

Problem 4.18 Suppose 8 dinner guests attend a dinner party and are seated at a circular table. Two of the guests are a couple that must sit together. If one of the seats is the "Head of the Table", how many ways are there to seat the guests? Hint: Be careful!

Problem 4.19 How many solutions to the equation $a+b+c+d+e=10$ are there if a, b, c, d, e are all non-negative integers?

Problem 4.20 Suppose you have 5 couples. How many ways are there to arrange them in a circle if all the couples stand next to each other?

4.3 Practice Questions

Problem 4.21 Let $A = \{1,2,3,4,5\}$, $B = \{4,5,6,7,8\}$. Verify the PIE formula for A, B. That is, calculate $n(A), n(B), n(A \cap B), n(A \cup B)$ directly and check the formula.

Problem 4.22 Write out the PIE formula for four sets A, B, C, D.

Problem 4.23 How many numbers between $1 - 100$ are a multiple of either $5, 7, 9$, or 11?

Problem 4.24 Suppose 4 people Albert, Bill, Charles, and Drew run a race. It was predicted that Albert would finish first, Bill second, and Charles third. How many outcomes of the race are there where all 3 of these predictions are wrong? (Note: We are not making any predictions about Drew.)

Problem 4.25 Suppose you have 8 different books. You want to use the books as gifts for 3 of you friends. How many ways are there to give out the gifts? Be nice: each friend gets at least one book! (Every gift is determined only by which books are given.)

Problem 4.26 Suppose you have a set $S = \{1,2,3,\ldots,40\}$. You want to choose A, B, C such that $A \cup B \cup C = S$ and $A \cap B \cap C = \emptyset$. How many ways are there to do so if $A \neq \emptyset$?

Problem 4.27 Suppose you have the numbers $\{0,1,2,3,4,5\}$. How many 6-digit numbers can be formed with 1 next to 0 or 2? (Use each number exactly once.) Caution: 012345 is not a 6 digit number so be careful!

Problem 4.28 There are objects in a bag defined by its color (red, blue) and shape (cube, ball). There are a total of 60 objects in the bag, none of which are blue cubes. 30 of the objects are red and 40 of the objects are balls. How many blue balls must be painted red in order to have a 20% increase in total number of red balls in the bag?

Problem 4.29 Suppose some friends go to a party. They each wear a coat. However, as they are leaving, they each randomly grab a coat. How many ways can the friends leave so that none of them have their own coat, if there are 5 friends?

Problem 4.30 Suppose you give out the 5 (current) ASOIAF books to 3 of your friends. Each friend gets at least one book. How many ways an you give out the books? Do the calculation using

(a) PIE.

(b) directly using cases.

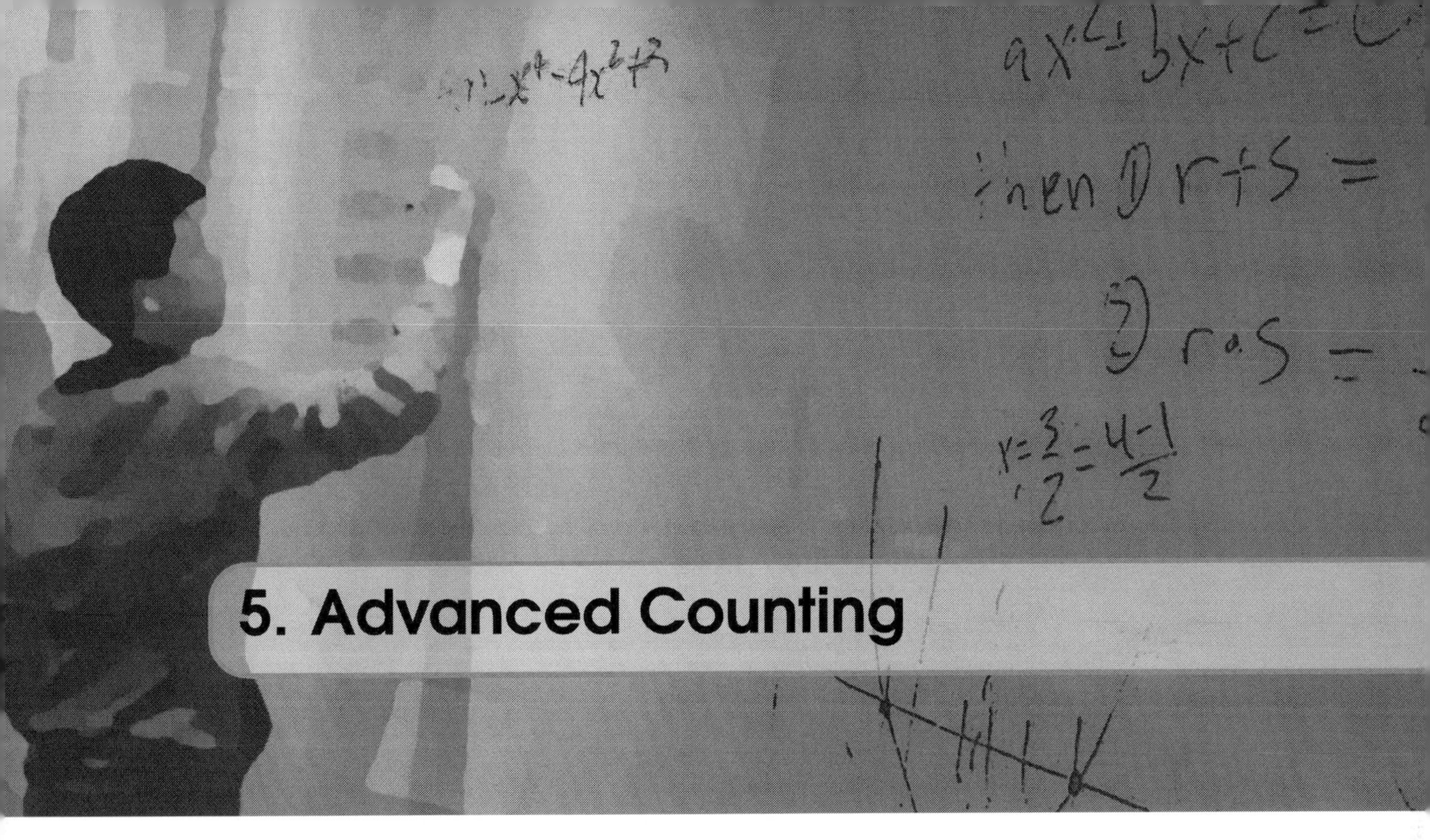

5. Advanced Counting

Basic Counting Rules

- Product Rule
- Sum Rule
- Permutations: Given a total of n objects, if you pick k of them in a specific order there are
$$\frac{n!}{(n-k)!}$$
ways to do so.
- Combinations: Given a total of n objects, if you pick k of them in no specific order there are
$$\binom{n}{k} = \frac{n!}{k! \cdot (n-k)!}$$
ways to do so.

Counting Methods

- Stars and Bars: If you put n identical balls into a k boxes (with no restrictions on how many balls go in each box) there are
$$\binom{n+k-1}{n} = \binom{n+k-1}{k-1} = \frac{(n+k-1)!}{n! \cdot (k-1)!}$$
ways to do so.
- Grouping: If some objects must be together, think of them as a single object/group when arranging.

- Spacing: If some objects must be separated, arrange the other objects first, and then place the objects in the spaces created by the others.

5.1 Example Questions

Problem 5.1 A train with 20 passengers must make 7 stops.

(a) How many ways are there for the passengers to get off the train at the stops?

(b) Repeat part (a) if we only care about the number of passengers getting off at each stop?

Problem 5.2 Suppose you have four black, four white, and four green balls. Assume balls of the same color are identical. How many ways are there to put all 12 balls into the 6 distinguishable boxes if

(a) if all 12 balls were black (that is, all identical)?

(b) if each box can have at most one ball of the same color? (Color matters!)

(c) if multiple balls of the same color can be in the same box?

Problem 5.3 The following problems are about a very smart grasshopper.

(a) Suppose a grasshopper sits at the center O of one corner of an 8×8 chessboard. At a given moment, it can jump to the center of any of the squares which have a common edge with the square where it currently sits, as long as the jump increases the distance between point O and the position of the grasshopper. How many ways are there for the grasshopper to reach the square at the opposite corner?

(b) A $8 \times 8 \times 8$ cube is formed of small unit cubes. A grasshopper "sits" at the center O of one corner cube. At a given moment, it can "jump" to the center of any of the cubes which have a common face with the cube where it currently sits, as long as the jump increases the distance between point O and the position of the grasshopper. How many ways are there for the grasshopper to reach the cube at the opposite corner?

Problem 5.4 Suppose you have 6 red cards and 20 black cards. Assume all the cards of the same color are identical. Deal the cards out in a line.

(a) How many different arrangements of the cards are there?

(b) Repeat part (a), if there must be at least 2 black cards between all the red cards.

Problem 5.5 Given positive integers $1, 2, 3, \ldots, 10$. Let a permutation of these numbers satisfy the requirement that, for each number, it is either (i) greater than all the numbers after it, or (ii) less than all the numbers after it. How many such permutations are there? For example, $1, 2, 3, 4, 5, 6, 7, 8, 9, 10$ and $10, 1, 2, 3, 4, 5, 6, 7, 8, 9$ both work.

Problem 5.6 Suppose you have 10 identical balls. How many ways are there to put them in 7 numbered boxes so that at least one of the boxes gets at least 4 balls?

Problem 5.7 Recall Problem 5.4 where we had 6 red cards and 20 black cards. Now suppose the 6 red cards are numbered and the cards are dealt out in a circle. (The black cards are still identical.)

(a) How many different arrangements of the cards are there?

(b) Repeat part (a), if there must be at least 2 black cards between all the red cards.

Problem 5.8 How many ways are there to write 201 as the sum of three non-negative integers (we care about the order of the numbers) if all three numbers must be different?

Problem 5.9 Consider the number 16000000.

(a) How many ways are there to represent it as the product of two factors, each divisible by 8, if we consider products that differ in the order of factors to be different?

(b) Repeat (a) if we do *not* care about the order of the factors.

Problem 5.10 Suppose you place 9 rings on the 3 mid fingers of your left hand (that is, not on your thumb or your pinky). How many different outcomes are possible if

(a) all the rings are identical, and no finger has more than 3 rings?

(b) all the rings are different, and no finger has more than 3 rings?

(c) all the rings are identical, and no finger has more than 8 rings?

(d) all the rings are different, and all the rings are on a single finger?

5.2 Quick Response Questions

Problem 5.11 We call a natural number "odd-looking" if all its digits are odd. How many four-digit odd-looking numbers are there?

Problem 5.12 How many ways are there to fill a Special Sport Lotto card? In this lotto you must predict the results of 13 hockey games, indicating either a victory for one of two teams or a draw.

(A) 3^{13}
(B) 13^3
(C) $\binom{13}{3}$
(D) $\frac{13!}{10!}$

Problem 5.13 The Hermetian alphabet consists of only three letters A, B, and C. A word in this language is an arbitrary sequence of no more than 3 letters. How many words does the Hermetian language contain?

Problem 5.14 There are six letters in the Kermetian language. A word is any sequence of six letters, some pair of which are the same. How many words are there in the Kermetian language?

(A) $\binom{6}{2} \cdot 6!$
(B) $\binom{6}{2} \cdot 5!$
(C) $6^6 - 6!$
(D) $6! - 4!$

Problem 5.15 There are 4 books on a shelf. How many ways are there to arrange some or all of them in a stack? (One single book can also be called a stack.)

Problem 5.16 Two couples and a family of three all line up for a photograph. How many photographs are possible with all the couples and families together?

Problem 5.17 A bookbinder must bind 12 identical books using red, green, or blue covers. In how many ways can this be done?

Problem 5.18 For how many positive integers $a \neq b$ does $a+b=80$?

Problem 5.19 How many rearrangements of 5 A's, 5 B's, and 5 C's are there?

(A) $\frac{15!}{3!}$
(B) $\frac{15!}{5!}$
(C) $\frac{15!}{10!}$
(D) $\frac{15!}{(5!)^3}$

Problem 5.20 How many ways are there to put 4 numbered balls into 3 numbered boxes so that each box has at least one ball?

5.3 Practice Questions

Problem 5.21 A train with 15 passengers must make 15 stops.

(a) How many ways are there for the passengers to get off the train at the stops, if not all the passengers get off at the same stop?

(b) Repeat part (a) if we only care about the number of passengers getting off at each stop?

Problem 5.22 Suppose you have four black, four white, and four green balls. How many ways are there to put the 12 balls into 6 distinguishable boxes if every color is put in at least 2 boxes?

Problem 5.23 Suppose an ant starts at the origin $(0,0)$. Ever step it takes is either $(1,1)$ or $(1,-1)$ (so it moves diagonally up and diagonally down).

(a) How many different ways can the ant move from the origin to $(20,0)$?

(b) Repeat part (a) if the ant makes a stop at $(10,0)$ along the way.

Problem 5.24 Suppose you have 8 numbered red cards and 20 identical black cards. How many ways are there to arrange the cards so that there is at least two black cards between each red card?

Problem 5.25 Suppose you have a sequence of 100 integers, and the first integer is 100. Every integer after the first is either one larger or smaller than the previous.

(a) How many different sequences of integers are possible?

(b) How many different possibilities are there for the last integer in the sequence?

Problem 5.26 Suppose you have 10 identical balls. How many ways are there to put them in 4 numbered boxes so that at least one of the boxes gets at least 3 balls?

(a) Show how to answer this question using the same method as in Example 5.6.

(b) Explain why an easier method can be used to solve this problem.

Problem 5.27 Suppose you have 6 numbered red cards and 20 numbered black cards. How many ways are there to arrange the cards in a circle if there must be at least 2 black cards between each red card?

Problem 5.28 How many ways are there to write 200 as the sum of three non-negative integers (we care about the order of the numbers) if all three numbers must be different?

Problem 5.29 Consider the number 40.

(a) How many ways are there to represent it as the product of three factors, if we consider products that differ in the order of factors to be different?

(b) Repeat (a) if we do not care about the order of the factors. Hint: List them out!

Problem 5.30 Suppose you place 2 different rings on 4 fingers (not your thumb) on your left hand. How many different outcomes are possible?

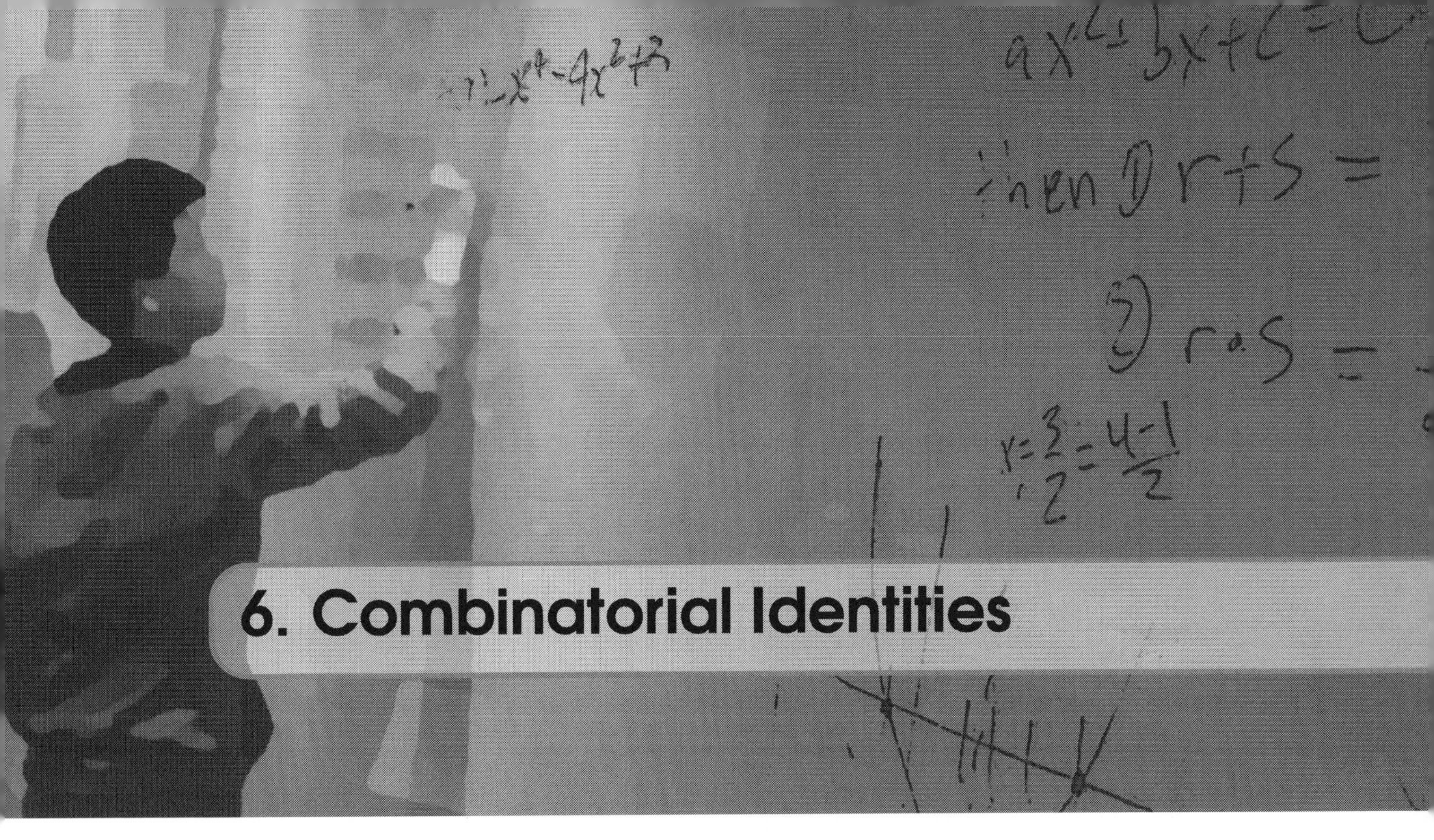

6. Combinatorial Identities

Binomial Coefficients

- Review:
  - Recall that $\binom{n}{k}$ was the number of way to choose a collection of k objects from a total of n (different) objects.
  - We have the formula $\binom{n}{k} = \dfrac{n!}{k! \cdot (n-k)!}$.
- Identities Involving Binomial Coefficients (to be proved later):
  - **Symmetry**: $\binom{n}{k} = \binom{n}{n-k}$ for $0 \le k \le n$.
  - **Pascal's Identity**: $\binom{n}{k} + \binom{n}{k+1} = \binom{n+1}{k+1}$.
  - **Reduction**: $k\binom{n}{k} = n\binom{n-1}{k-1}$ for $0 \le k \le n$.
  - **Hockey Stick Identity**: For n, r positive integers with $n > r$,

$$\binom{r}{r} + \binom{r+1}{r} + \binom{r+2}{r} + \cdots + \binom{n}{r} = \binom{n+1}{r+1},$$

    or, using summation notation, $\sum_{k=r}^{n} \binom{k}{r} = \binom{n+1}{r+1}$.
- **Relationship with Pascal's Triangle**: It is useful to think of many of these

identities in terms of Pascal's Triangle.

$$\binom{0}{0}$$
$$\binom{1}{0} \quad \binom{1}{1}$$
$$\binom{2}{0} \quad \binom{2}{1} \quad \binom{2}{2}$$
$$\binom{3}{0} \quad \binom{3}{1} \quad \binom{3}{2} \quad \binom{3}{3}$$

$$1$$
$$1 \quad 1$$
$$1 \quad 2 \quad 1$$
$$1 \quad 3 \quad 3 \quad 1$$

Combinatorial Proof

- A *combinatorial proof* is a technique to prove an identity by solving a single counting question using two different arguments. The identity is thus proved as both arguments must lead to the same answer.
- For example, consider $\binom{n}{k}$, which was defined earlier in the handout as the number of ways to choose a collection of k objects from a total of n (different) objects. Using this definition, we can then use a combinatorial proof to *prove* $\binom{n}{k} = \frac{n!}{k! \cdot (n-k)!}$: First order all the objects ($n!$ ways). Then we don't care about the last $(n-k)$ objects, so we divide by $(n-k)!$. This leaves us with our k objects (in order) so we also divide by $k!$.
- Many different identities and formulas in combinatorics can be shown using a combinatorial proof.

The Binomial Theorem and Multinomial Theorem

- **Binomial Theorem**: Let n be a positive integer, then

$$(a+b)^n = \binom{n}{0}a^n + \binom{n}{1}a^{n-1}b + \cdots + \binom{n}{k}a^{n-k}b^k + \cdots + \binom{n}{n-1}ab^{n-1} + \binom{n}{n}b^n,$$

or, using summation notation, $(a+b)^n = \sum_{k=0}^{n} \binom{n}{k} a^{n-k}b^k..$

- **Binomial Theorem Alternative Version**: Let n be a positive integer. Then the coefficient of $a^{n-k}b^k$ in $(a+b)^n$ is $\binom{n}{k}$.
- **Multinomial Theorem**: Let n be a positive integer. Then the coefficient of $a_1^{k_1}a_2^{k_2}\cdots a_j^{k_j}$ (where $k_1+k_2+\cdots+k_j=n$) in $(a_1+a_2+\cdots+a_j)^n$ is $\frac{n!}{(k_1!)\cdot(k_2!)\cdots(k_j!)}$.

6.1 Example Questions

Problem 6.1 Prove the following algebraically:

(a) The symmetry formula.

(b) Pascal's Identity.

(c) The reduction formula.

Problem 6.2 Calculate the following. The identities may be useful in simplifying your answer.

(a) You have 10 friends and you give out (identical) copies of Monopoly to some of them. If no one gets more than one copy, how many ways are there to give out games to 5 or 6 of your friends?

(b) There are 4 types of presents: (i) Monopoly, (ii) a basketball, (iii) a t-shirt, and (iv) chocolate cookies. (You have multiple identical copies of each of the 4 types.) You decide to create a gift collection (containing some or all of the presents above). How many ways are there to create the collection if it contains between 0 and 7 presents?

Problem 6.3 Prove the following using a combinatorial proof.

(a) The symmetry formula.

(b) The reduction formula. Hint: Consider a committee with a president.

Problem 6.4 Answer each of the questions below in two different ways. Then expand the arguments to prove the remaining identities about binomial coefficients.

(a) Suppose you have one friend named Gwen, as well as 10 other friends. You want to invite a group of 5 friends out to dinner. How many different groups of friends could you invite?

(b) Consider a city grid below:

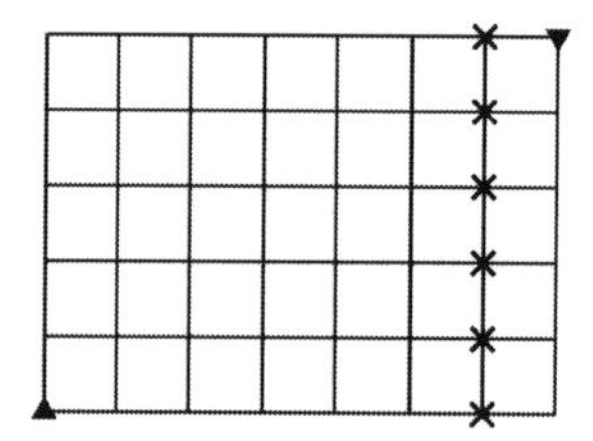

How many paths are there from the lower left triangle to the upper right triangle? Hint: The $\times$'s are in the diagram for a reason!

Problem 6.5 Use the following questions as a guide, give a proof of the binomial theorem.

(a) How many words can you form with $n-k$ $A's$ and k B's?

(b) Argue that, when expanding $(a+b)^n$, you get the sum of all words of length n consisting of a's and b's. For example, $(a+b)^2 = aa+ab+ba+bb = a^2+2ab+b^2$.

Problem 6.6 Simplify the following

(a) $\binom{n}{0}+\binom{n}{1}+\binom{n}{2}+\cdots+\binom{n}{n}$.

(b) $\binom{n}{0}-\binom{n}{1}+\binom{n}{2}+\cdots+(-1)^n\binom{n}{n}$.

Problem 6.7 Simplify $\sum_{k=0}^{n}\frac{1}{k+1}\binom{n}{k}$.

Problem 6.8 Find the coefficient of

(a) x^4y^5 in $(x+y)^9$.

(b) x^4 in $(x+3)^9$.

Problem 6.9 Find the constant term in the expansion of $\left(\sqrt{x}+\frac{1}{\sqrt{x}}-2\right)^5$.

Problem 6.10 Prove that $\sum_{j=0}^{n}\binom{m}{j}\cdot\binom{n-m}{k-j}=\binom{n}{k}$. Hint: $(1+x)^n=(1+x)^m(1+x)^{n-m}$.

6.2 Quick Response Questions

Problem 6.11 Suppose a shop has 6 types of cookies and want to buy 8 cookies. How many ways can you place an order of cookies if you are allowed to order multiple cookies of the same type?

Problem 6.12 How many ways are there to buy the cookies if you want 3 cookies each of 2 types of cookies, and 2 cookies of a third type?

Problem 6.13 How many ways can you order the 8 cookies if you want an even number of each cookie type? (Remember 0 is an even number.)

Problem 6.14 Write $\dfrac{6!}{2!\cdot 2!\cdot 2!}$ in terms of binomial coefficients.

(A) $\binom{6}{2}\cdot\binom{6}{2}\cdot\binom{6}{2}$
(B) $\binom{6}{4}\cdot\binom{6}{2}$
(C) $\binom{6}{4}\cdot\binom{4}{2}$
(D) $\binom{6}{2}\cdot\binom{6}{2}$

Problem 6.15 The ratio of $\binom{10}{7}:\binom{9}{6}$ in reduced form is $a:b$. What is $a+b$?. Hint: You can use one of the identities!

Problem 6.16 Calculate $\binom{5}{0}+\binom{6}{1}+\binom{7}{2}+\binom{8}{3}$. Hint: This is a disguised version of one of the identities.

Problem 6.17 Calculate $\binom{10}{0} - \binom{10}{1} + \binom{10}{2} - \cdots + \binom{10}{10}$.

Problem 6.18 Calculate $2 \cdot \binom{10}{0} + 2 \cdot \binom{10}{2} + 2 \cdot \binom{10}{4} + \cdots + 2 \cdot \binom{10}{10}$.

Problem 6.19 Find the coefficient of x^3 in $(x-5)^5$.

Problem 6.20 Find the coefficient of $x^2y^3z^5$ in $(x+y+z)^{10}$.

6.3 Practice Questions

Problem 6.21 Prove the following identities algebraically:

(a) $\binom{n}{k} \cdot \binom{n-k}{j} = \binom{n}{j} \cdot \binom{n-j}{k}$.

(b) $k\binom{n}{k} = (n+1-k)\binom{n}{k-1}$.

Problem 6.22 Prove the identity

$$\binom{n}{k} \cdot \binom{n-k}{j} = \binom{n}{j} \cdot \binom{n-j}{k}$$

using a combinatorial proof.

Problem 6.23 Prove the identity

$$k\binom{n}{k} = (n+1-k)\binom{n}{k-1}$$

using a combinatorial proof.

Problem 6.24 Use the hockey stick identity to prove that $1+2+3+\cdots+n = \frac{n(n+1)}{2}$.
Hint: First write everything in terms of binomial coefficients.

Problem 6.25 Prove the form of the multinomial theorem given above.

Problem 6.26 Simplify $\binom{n}{0} + 2\cdot\binom{n}{1} + 4\cdot\binom{n}{2} + \cdots + 2^n\cdot\binom{n}{n}$.

Problem 6.27 Simplify $\sum_{k=1}^{n} k\binom{n}{k}$.

Problem 6.28 Find all the terms (including their coefficient) in $(x+y+z+4)^9$ which contain x^4y^2. For example: the terms containing x in $(x+y+1)^2$ are $2x$ and $2xy$.

Problem 6.29 Find the coefficient of x^{11} in $(x^5+x+2)^{10}$. Caution: $11 = 10+1$ but also $11 = 5+6$.

Problem 6.30 Prove that $\binom{4n+2}{0} - \binom{4n+2}{1} + \binom{4n+2}{2} - \binom{4n+2}{3} + - \cdots + \binom{4n+2}{2n} = \frac{1}{2}\binom{4n+2}{2n+1}$.

7. Recurrence Relations

Sequences

- Recall a *sequence* is a list of numbers (either finite or infinite) $a_0, a_1, a_2, \ldots$. Caution: Note we're starting with a_0!
- An *arithmetic* sequence has formula $a_n = a_0 + k \cdot n$ where a_0 is the starting value and k is the common difference. Alternatively, we can give a *recursive* definition: $a_{n+1} = a_n + k$.
- An *geometric* sequence has formula $a_n = a_0 \cdot r^n$ where a_0 is the starting value and r is the common ratio. Alternatively, we can give a *recursive* definition: $a_{n+1} = r \cdot a_n$.
- Recursive formulas are often useful for understanding a sequence and can sometimes (but not always!) be turned into a general formula.

Functions

- Recall that a function is just a mapping between two sets A and B.
- Functions are often given by equations, but this is not necessary. For example, "the first letter of your first name" can be thought of as a function from the set of all names to the set $\{A, B, C, \ldots, Z\}$ of letters of the alphabet.
- A function is *injective* or *one-to-one* if every output has exactly one input. For example, $f(x) = x$ is injective, but $f(x) = x^2$ is not.
- A function is *surjective* or *onto* if every value in the target set is the output of some input. That is, the range of the function is the same as the target set. As functions

from the real numbers to the real numbers $f(x) = x$ is surjective, but $f(x) = x^2$ is not surjective.
- If a function is injective and surjective it is called a *bijection*.
- In other words, given two sets A, B, a bijection from A to B is a one-to-one correspondence between members of A and members of B. That is, every element in a "matches" up with exactly one element in "B".

Combinatorial Proof and Bijections

- We use bijections all the time in combinatorics (even if we don't know it). For example, there is a natural bijection between the outcomes of flipping a coin 4 times and words of length 4 made up of only the letters H, T.
- Clearly, if there is a bijection between A and B, then A and B have the same size. This can be useful in solving some counting questions.
- Therefore, a combinatorial proof is more formally understood as describing a bijection between two sets (which therefore have the same size).
- Combinatorial proofs can often be used to prove recurrences.
- For example, Pascal's Identity: $\binom{n+1}{k+1} = \binom{n}{k} + \binom{n}{k+1}$ can be thought of as a recurrence for binomial coefficients, and we saw a combinatorial proof last time.

7.1 Example Questions

Problem 7.1 Turn each recursive definition of a sequence given below into a general formula.

(a) $a_0 = -4, a_{n+1} = a_n + 2$.

(b) $a_0 = 8, a_{n+1} = a_n/2$.

(c) $a_0 = 1, a_{n+1} = (n+1) \cdot a_n$.

(d) $a_0 = 0, a_{n+1} = a_n + n + 1$.

Problem 7.2 Find recursive definitions of the following sequences. Writing out a few small examples might help! You do not need to prove your answers.

(a) Let F_n (for $n \geq 1$) denote the number of ways to write $n-1$ as the sum of 1's and 2's. For example, $3 = 1+1+1 = 2+1 = 1+2$ so $F_4 = 3$.

(b) Let a_n (for $n \geq 1$) denote the number of ways to invite a group of your friends (assume there are n total friends) if you invite at least one friend but not all the friends.

Problem 7.3 Let $A = \{1,2,3,4,\ldots,10\}$ and $B = \{1,2,3,4\}$. Answer each of the following. Explain how you have already answered this type of question in the past.

(a) How many total functions are there from A to B? from B to A?

(b) How many bijections are there from A to B?

(c) How many injections are there from B to A?

(d) How many surjections are there from A to B?

Problem 7.4 Let $A = \{1,2,3,4,5\}$.

(a) Give an example of a bijection between subsets of A of size two and subsets of A of size 3.

(b) How many such bijections are there in part (a)?

(c) Find a bijection between subsets of A of size 2 and two numbers chosen (repetition allowed) from $\{1,2,3,4\}$.

Problem 7.5 Complete the following proofs combinatorially.

(a) Let $a_n =$ the number of subsets of $\{1,2,\ldots,n\}$. (So $a_n = 2^n$.) Prove combinatorially that $a_n = 2 \cdot a_{n-1}$.

(b) Let $a_n =$ the number of permutations of $\{1,2,\ldots,n\}$. (So $a_n = n!$.) Prove combinatorially that $a_n = n \cdot a_{n-1}$. Hint: If you permute $\{1,2,\ldots,n-1\}$ how many "spaces" are available for n?

Problem 7.6 Carol and Tom are getting married. They are going through their friends and deciding if (i) the friend is invited to the wedding and the reception, (ii) the friend not invited to the wedding but invited to the reception, or (iii) not invited at all. Assume they invite at least one of their friends to the reception and let a_n denote the number of ways they can give out invitations to n friends.

(a) Give a general formula for a_n.

(b) Give a recursive formula for a_n. Prove your answer combinatorially.

Problem 7.7 Three identical standard dice are thrown. How many possible rolls are there? (That is, we only care about how many of each number are rolled.)

(a) Give a solution using stars and bars.

(b) Give a solution using a bijection. Hint: Consider your answer to part (a).

Problem 7.8 In Problem 7.2 we saw the *Fibonacci* sequence, with recursive definition $F_n = F_{n-1} + F_{n-2}$, and $F_1 = F_2 = 1$ (or sometime $F_0 = 0$). We also saw that $F_n =$ the number of ways to write $n-1$ as a sum of 1's and 2's. Use this characterization to prove the recurrence $F_n = F_{n-1} + F_{n-2}$.

Problem 7.9 Suppose n friends go to a party. They each wear a coat. However, as they are leaving, they each randomly grab a coat. Let $a_n =$ the number of ways the friends can leave so that *none* of them have their own coat.

(a) $a_0 = 1$ by convention. Find a_1, a_2.

(b) Prove combinatorially that $a_n = (n-1) \cdot (a_{n-1} + a_{n-2})$. Hint: With n people, the first person can choose one of $n-1$ hats. Then consider two cases.

Problem 7.10 Let F_n denote the nth Fibonacci number. Prove

$$F_1^2 + F_2^2 + \cdots + F_n^2 = F_n F_{n+1}$$

geometrically. Interpret F_k^2 as the area of a square with side length F_k and $F_n \cdot F_{n+1}$ as the area of an F_n by F_{n+1} rectangle.

7.2 Quick Response Questions

Problem 7.11 Find the coefficient of x^5y^8 in $(x+y)^{13}$.

Problem 7.12 Find the coefficient of x^8 in $(x+3)^{12}$.

Problem 7.13 Find the coefficient of $x^5y^4z^2$ in $(x+y+z)^{11}$.

Problem 7.14 Calculate $\binom{7}{0}+2\cdot\binom{7}{1}+4\cdot\binom{7}{2}+\cdots+128\cdot\binom{7}{7}$.

Problem 7.15 Calculate $\binom{8}{0}-2\cdot\binom{8}{1}+4\cdot\binom{8}{2}-8\cdot\binom{8}{3}+-\cdots+256\cdot\binom{8}{8}$.

Problem 7.16 Write out the first few terms of the recurrence relation $a_0=2$, $a_{n+1}=\frac{1}{1-a_n}$ until you find a pattern. This pattern repeats every K terms. What is K?

Problem 7.17 Which of the options below gives a general formula for a sequence with recursive definition $a_0=4, a_{n+1}=3\cdot a_n-2$?

(A) $a_n=3^n+3$
(B) $a_n=3^{n+1}+3$
(C) $a_n=6n+4$
(D) $a_n=3\cdot 3^n+1$

Problem 7.18 Suppose you pick an ordered collection (repeats allowed) of three numbers from the set $\{1,2,3,4,5\}$. How many different collections are possible?

Problem 7.19 Suppose you pick an unordered collection (repeats allowed) of six numbers from the set $\{1,2,3,4,5,6,7,8\}$. How many different collections are possible?

Problem 7.20 Suppose n friends go to a party. They each wear a coat. However, as they are leaving, they each randomly grab a coat. Let $a_n =$ the number of ways the friends can leave so that *none* of them have their own coat. In Problem 7.9 we gave a recursive formula for a_n.

Calculate a_4, and verify the calculation by listing out all of the outcomes. Let the people $1,2,3,4$ have hats A,B,C,D. Thus, one example is *DCBA* (person one getting coat D, person two had C, etc.). (Input the value of a_4 as your answer.)

7.3 Practice Questions

Problem 7.21 Write a general formula for a sequence with recursive definition $a_0 = 1, a_{n+1} = \sum_{k=0}^{n} a_k$. Hint: It is okay to write a formula that only works for $n \geq 2$.

Problem 7.22 Write a recursive definition for the sequence $a_n = \frac{n(2n+1)(2n-1)}{3}$ for $n \geq 1$. Hint: Write out the first few terms!

Problem 7.23 Let $A = \{1,2,3,4,5\}$ and B be the set of all subsets (of any size) of A. How many injections are there from A to B?

Problem 7.24 Find a "natural" bijection between the set $A = \{0,1,2,3\cdots,2^n - 1\}$ and the set $P(B)$ of all subsets of $B = \{1,2,3,\cdots,n\}$. Hint: think of the elements of A as being written in binary.

Problem 7.25 Prove combinatorially that the number of ways to put n identical balls into k different boxes is the same as the number of non-negative solutions to $a_1 + a_2 + \cdots + a_k = n$. That is, write out a bijection between outcomes in the first case and outcomes in the second case. Note: This is easy, but it is good practice to carefully write out the bijection.

Problem 7.26 Let $a_n =$ the number of ways to invite a group of friends (from n total) if you invite at least one but not all your friends. Recall in Problem 7.2 we showed $a_n = 2a_{n-1} + 2$.

(a) Find a (non-recursive) formula for a_n.

(b) Prove the recursive formula $a_n = 2a_{n-1} + 2$ combinatorially.

Problem 7.27 Suppose 3 numbers are chosen from the set $\{1, 2, \ldots, 7\}$. In how many ways can this be done such that the chosen subset has at least one pair of neighbors? Hint: Use complementary counting.

Problem 7.28 How many ways are there to tile (that is, cover completely with no overlaps) a 2×6 grid with 1×2 dominoes? Hint: Find a pattern. Extra: Prove your pattern combinatorially!

Problem 7.29 Suppose n friends go to a party. They each wear a coat. However, as they are leaving, they each randomly grab a coat. Let $a_n =$ the number of ways the friends can leave so that none of them have their own coat.

(a) Recall that we've already solved this type of problem in previous weeks using PIE. Explain the formula: $a_n = \sum_{k=0}^{n} (-1)^k \cdot \frac{n!}{k!}$.

(b) Verify that the formula given in part (a) and the recursive formula in Problem 7.9 give the same answer for a_8.

Problem 7.30 Prove the identity $\sum_{k=1}^{n} F_k = F_{n+2} - 1$ combinatorially. (F_n denotes the Fibonacci sequence.) Hint: Recall F_n is the number of ways to write $n-1$ as a sum of 1's and 2's. Consider the location of the first 2.

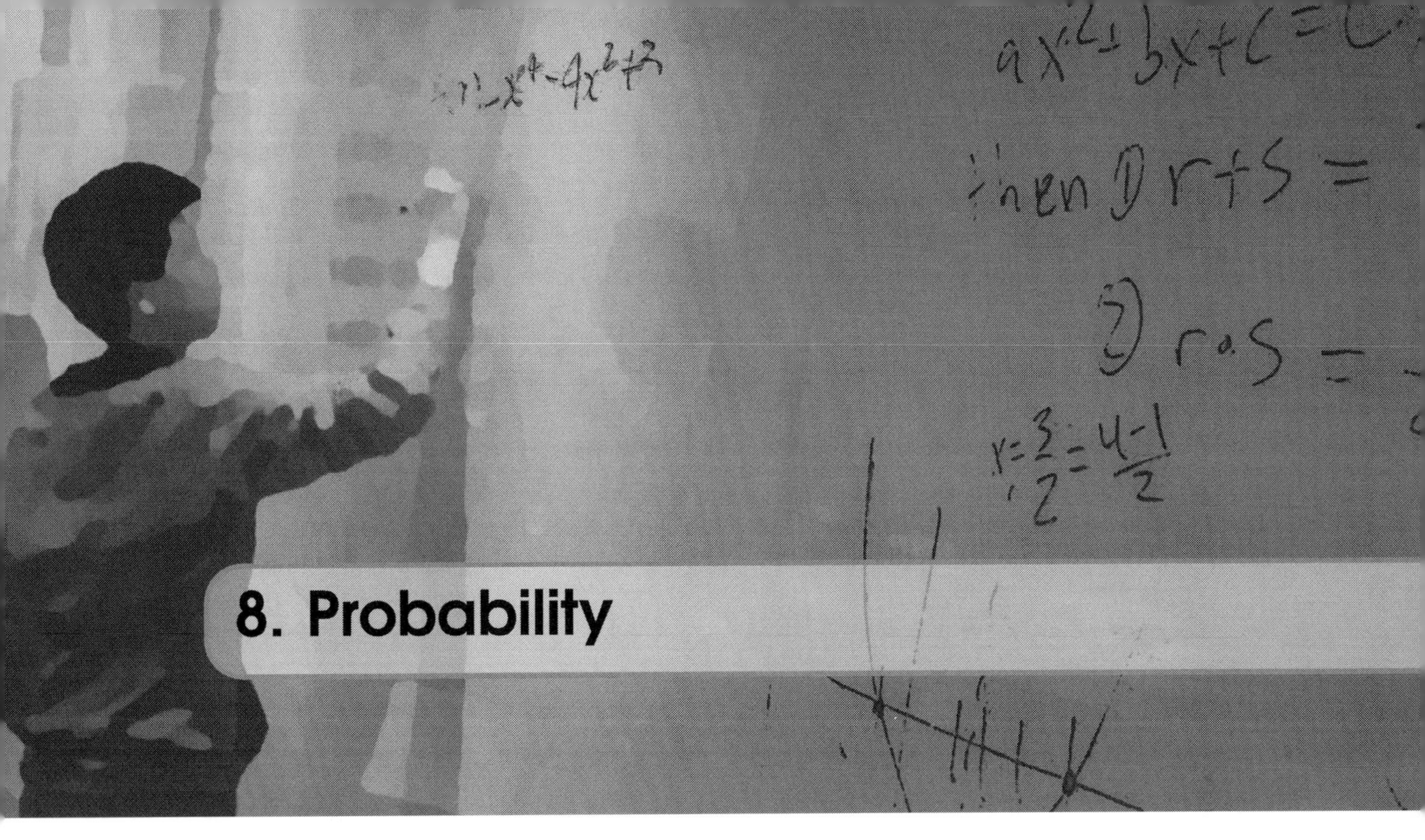

8. Probability

Review of Sets and Notation

- Recall that a *set* is an unordered collection of objects, without repetitions. We call the members of a set its *elements*.
- For example, $\{1,4,5\}$ is a set of 3 numbers. We have

$$\{1,4,5\} = \{4,1,5\} = \{1,1,4,5\}.$$

- Recall a set B is a *subset* of A, written $B \subseteq A$ if every element of B is an element of A.
- In probability, we call the set of all possible outcomes of an experiment the *sample space*, and denote it by Ω (the capital Greek letter omega).
- We will call a subset A of Ω (written $A \subseteq \Omega$) an *event*.
- In probability, the complement of an event A is the collection of all elements of Ω not in A, denoted A^c.
- The notation $A \cup B$ denotes the elements in either A or B (or both). This is called the *union* of A and B.
- The notation $A \cap B$ denotes the elements in both A and B. This is called the *intersection* of A and B.
- The *empty set* is the set with no elements. It is denoted by $\emptyset$ or sometimes $\{\}$.
- If a set A is finite, we use the notation $n(A)$ to denote the number of elements in A.
- Note: Sets can contain more than just numbers. For example, we could have the set of all states in the US, or the set of all words starting with the letter A, etc.

Probability (Classical Model)

- Suppose Ω is a finite sample space and every outcome in Ω is equally likely. If $A \subseteq \Omega$, then
$$\text{the probability of } A = P(A) = \frac{\text{want}}{\text{total}} = \frac{n(A)}{n(\Omega)}.$$
- For example, if we flip a fair coin twice, then $\{HH, HT, TH, TT\}$ is a sample space where every outcome is equally likely. However, the sample space {two heads, one head and one tail, two tails} is a sample space but each outcome is *not* equally likely.

Geometric Probability

- Remember that our formal formula for probability only works with a finite sample space.
- Infinite sample spaces are often difficult to work with, but one fairly easy case is when Ω is a geometric shape.
- Suppose that Ω is a geometric shape (line, square, circle, cube, sphere, etc.) and $A \subseteq \Omega$. Then
$$\begin{aligned} P(A) = \frac{\text{want}}{\text{total}} &= \frac{\text{want (in terms of length)}}{\text{total (in terms of length)}} \text{ OR} \\ &= \frac{\text{want (in terms of area)}}{\text{total (in terms of area)}} \text{ OR} \\ &= \frac{\text{want (in terms of volume)}}{\text{total (in terms of volume)}}. \end{aligned}$$
- For problems involving geometric probability, it is often useful to draw a diagram.

Axioms (Rules) of Probability

- Suppose that Ω is a sample space and $A, B \subseteq \Omega$ are events.
  Pr1. $P(A) \geq 0$.
  Pr2. $P(\Omega) = 1$.
  Pr3. If A and B are *disjoint* (that is, $A \cap B = \emptyset$), then $P(A \cup B) = P(A) + P(B)$.
- Additional properties which can be proven from the axioms.
  Pr4. $P(\emptyset) = 0$.
  Pr5. $0 \leq P(A) \leq 1$.
  Pr6. $P(A^c) = 1 - P(A)$.
  Pr7. $P(A \cup B) = P(A) + P(B) - P(A \cap B)$.

8.1 Example Questions

Problem 8.1 Suppose you roll 2 fair six-sided dice. Let A be the event that the first die is a 6, B the event that the sum of the two rolls is 7, and C be the event that the sum is even.

(a) Calculate $P(A)$, $P(B)$, $P(C)$.

(b) Find $P(A \cap B)$.

(c) Find $P(B \cup C^c)$.

(d) Find $P(A \cap C)$.

Problem 8.2 Suppose you flip a fair coin 8 times.

(a) What is the probability you get 5 heads?

(b) What is the probability you get an equal number of heads and tails?

(c) What is the probability you get more heads than tails?

Problem 8.3 Suppose you randomly pick a number on the number line between -3 and 5.

(a) What is the probability the number is positive?

(b) What is the probability that the number squared is greater than 4?

Problem 8.4 Suppose Jack and Jill both randomly come to school between 9 AM and 1 PM.

(a) What is the probability that Jack comes to school before 10 AM?

(b) What is the probability Jack and Jill both come to school before 11 AM?

(c) What is the probability Jack comes to school before Jill?

(d) What is the probability Jack and Jill come to school within 1 hour of each other?

Problem 8.5 Suppose I pick a number from $\Omega = \{1,2,3,4,5,6,7,8\}$. However, each number is not equally likely to be chosen: the probability I pick 1, 3, or 5 is .15, the probability I pick 2 or 4 is .05, the probability I pick 6 is .25, and the probability I pick 7 is the same as the probability I pick 8.

(a) Find $P(7)$ and $P(8)$.

(b) Find the probability I pick an even number.

(c) Find the probability I pick a prime number.

Problem 8.6 Suppose $A,B \subseteq \Omega$ such that $P(A) = .4, P(B) = .4$ and $P(A \cup B^c) = .8$. Find $P(A^c \cap B^c)$. Hint: Draw a Venn Diagram!

Problem 8.7 Suppose you are dealt a five card hand from a standard deck of 52 cards (with 13 ranks $2,3,\cdots,10,J,Q,K,A$ and 4 suits: hearts, diamonds, clubs, and spades). Find the probability of:

(a) a full house (3 cards of one rank, 2 cards of another).

(b) exactly two pair. (That is, not four of a kind or a full house.)

(c) a flush (all cards of the same suit).

(d) a straight (all 5 ranks in a row, so $A,2,3,4,5$ up to $10,J,Q,K,A$).

Problem 8.8 Suppose you have line segments of length $1,a,b$, where a,b are real numbers chosen randomly between 0 and 2. What is the probability you can form a triangle using the three line segments.

Problem 8.9 A bag contains 6 red, 5 green, and 4 yellow balls. You pick 4 balls at once (so in no particular order).

(a) What is the probability you get 3 red, 1 green, and 1 yellow ball?

(b) What is the probability you get 0 yellow balls?

(c) What is the probability you get all 4 balls of the same color?

(d) What is the probability you do not get all red and green balls?

Problem 8.10 Suppose Alice, Bob, and Charlie plan to meet for dinner. They each randomly show up to dinner between 5:00 and 6:00 PM. Alice will wait for Bob and Charlie to show up. Bob will wait for Charlie but not Alice. Charlie will not wait. What is the probability they have dinner together?

8.2 Quick Response Questions

Problem 8.11 Calculate $\sum_{k=4}^{10} \binom{k}{4}$. Input your answer as an integer.

Problem 8.12 Calculate the coefficient of x^7 in $(x-3)^{10}$.

Problem 8.13 Calculate the coefficient of x^4y^2z in $(x+y+z)^7$.

Problem 8.14 Which of the following is a general formula for the sequence with recursive definition $a_0 = 2$, $a_1 = 1$, $a_n = 2a_{n-1} - a_{n-2}$.

(A) $a_n = 2 \cdot 2^{-n}$
(B) $a_n = 2 - n$
(C) $a_n = 2 - n^2$
(D) $a_n = n^2 - 2n + 2$

Problem 8.15 Suppose you roll a fair six-sided die twice. The probability the sum is 4 can be written as $\frac{P}{36}$. What is P?

Problem 8.16 Suppose you are dealt 2 cards from a deck of cards. The probability you get two cards of the same suit can be written as $\frac{P}{Q}$ for positive integers P, Q with $\gcd(P, Q) = 1$. What is $P + Q$?

Problem 8.17 Suppose $\Omega = \{1,2,3,4\}$ and $P(1) = 0.1, P(3) = 0.3$ and $P(2) = P(4)$. Find $P(4)$.

Problem 8.18 Suppose $P(A) = 0.6$, $P(B) = 0.5$ and $P(A^c \cap B^c) = 0.3$. Find $P(A \cap B)$.

Problem 8.19 Suppose you randomly pick a real number c between 0 and 7. The probability you can create a triangle with side lengths $2,3,c$ can be written as $\frac{P}{Q}$ for positive integers P, Q with $\gcd(P,Q) = 1$. What is $Q - P$?

Problem 8.20 Suppose you write a computer program that randomly generates two real (floating point) numbers x and y between 0 and 5. The probability that $x + y \leq 4$ is $K\%$, where K is rounded to the nearest integer if necessary. What is K?

8.3 Practice Questions

Problem 8.21 Suppose you have a fair 5-sided die. You roll it 3 times. What is the probability that the sum is odd?

Problem 8.22 Suppose you flip a fair coin 8 times.

(a) What is the probability you get at least 3 heads?

(b) What is the probability you get a string of at exactly 5 or 6 heads in a row and the rest tails in the 8 flips?

Problem 8.23 Suppose you randomly pick a number (call it x) between -5 and 5. Find the probability that $|x-1| > 2$.

Problem 8.24 After school, Jack and Jill plan to meet up for dinner. They both randomly arrive between 5:00 and 6:00 PM. Jill will wait up to 15 minutes for Jack. Jack will wait up to 30 minutes for Jill, but will not wait at all after 5:45. What is the probability they have dinner together?

Problem 8.25 Suppose $\Omega = \{1,2,3,4\}$, and the probability of each number in Ω is half the probability of the previous value (so $P(3) = P(2)/2$ for example).

(a) Find $P(4)$.

(b) Find the probability of getting an odd number.

Problem 8.26 Let $A, B \subseteq \Omega$ be events. Suppose $P(A) = P(B) = .3$ and $P(A^c \cap B^c) = .6$. Find $P(A \cap B)$. Hint: draw a Venn diagram.

Problem 8.27 Suppose you are dealt 5 cards from a deck of cards. Find the probability of:

(a) a four of a kind.

(b) exactly one pair. (That is, not two pair or three of a kind, etc.)

Problem 8.28 Suppose you pick a point randomly in a hexagon with side length 2. What is the probability that the distance between the point chosen and a vertex is less than 1?

Problem 8.29 Pick 4 balls at once from 7 red, 6 green, and 5 yellow. What is the probability you get at least 1 red, and at least 1 green and at least 1 yellow ball? Hint: Be careful here!

Problem 8.30 Suppose you randomly pick a real number x between 0 and 3. What is the probability you can create a triangle with side lengths $1, x, x$?

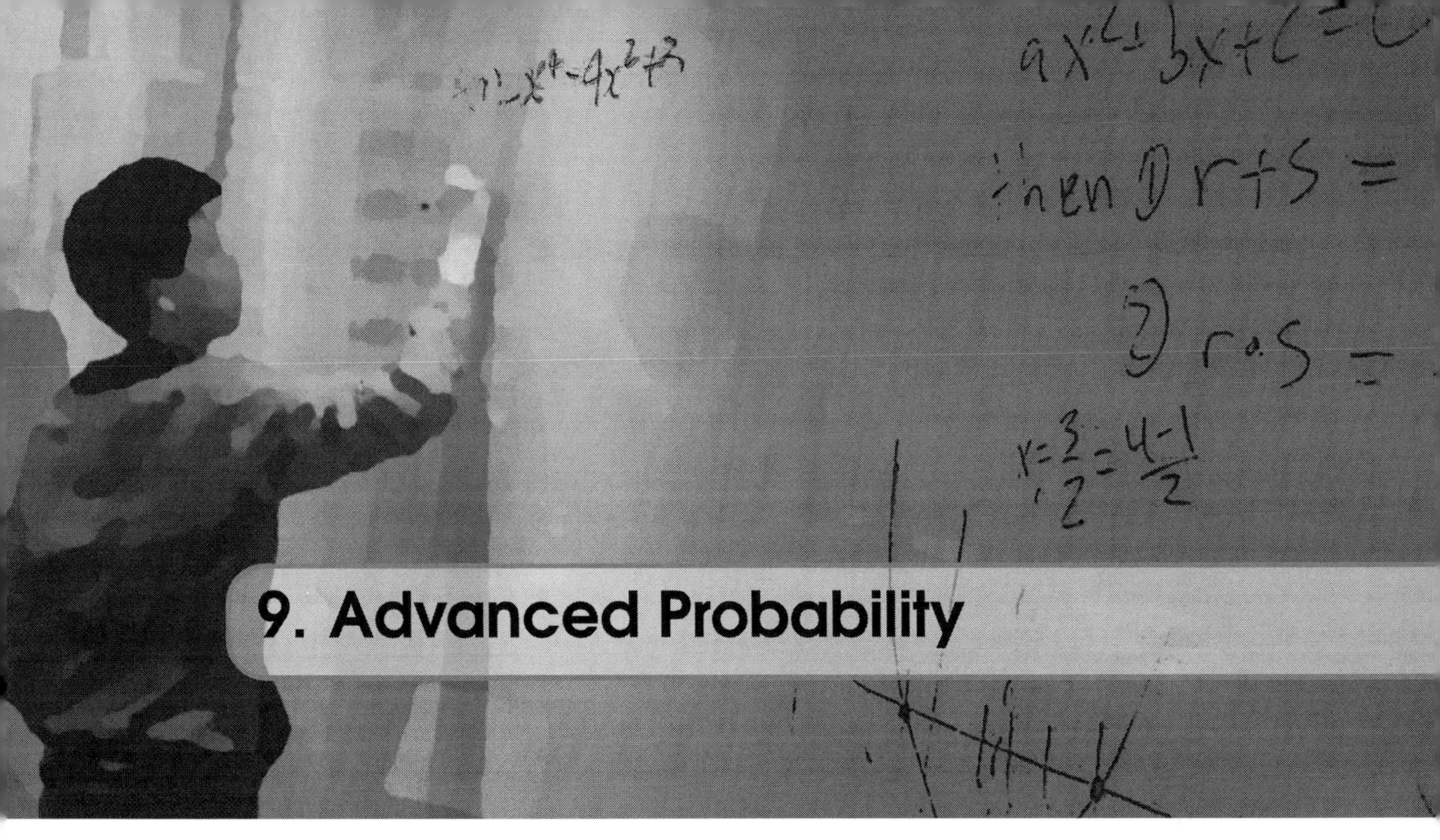

9. Advanced Probability

Conditional Probability

- Let A and B be two events. The *conditional probability* of A given B, written $P(A|B)$, is the probability of A knowing (for sure) that B happened.
- For example, while the probability of rolling a 6 with a standard fair die is $1/6$, the probability of rolling a 6 given that the roll is even is $1/3$.
- Suppose Ω is a finite sample space and every outcome in Ω is equally likely. If $A, B \subseteq \Omega$, then $P(A|B) = \dfrac{P(A \cap B)}{P(B)}$.

Independence

- Two events A, B are *independent* if knowing one happened doesn't affect the probability of the other.
- In mathematical terms, A, B are independent if $P(A|B) = P(A)$ AND $P(B|A) = P(B)$.
- Equivalently, A, B are independent if and only if $P(A \cap B) = P(A)P(B)$.
- Some examples of independent events: flipping coins, rolling dice, picking cards/balls with replacement/repetition, etc.

Law of Total Probability

- Events $B_1, \ldots, B_n$ are *pairwise disjoint* if $B_i \cap B_j = \emptyset$ for $i \neq j$.
- If $B_1, \ldots, B_n$ are pairwise disjoint and $B_1 \cup \cdots \cup B_n = \Omega$ we say $B_1, \ldots, B_n$ *partition* Ω.

- Suppose $A \subseteq \Omega$ is an event and $B_1, \ldots, B_n$ partition Ω. Then

$$P(A) = P(B_1) \cdot P(A|B_1) + \cdots + P(B_n) \cdot P(A|B_n) = \sum_{i=1}^{n} P(B_i) \cdot P(A|B_i).$$

Bayes' Theorem

- As a special case, A and A^c form a partition of Ω.
- The usual formulation of Bayes' Rule is

$$P(A|B) = \frac{P(A) \cdot P(B|A)}{P(A) \cdot P(B|A) + P(A^c) \cdot P(B|A^c)}.$$

- The formulation of the Bayes' Theorem can be generalized (using other partitions of Ω), but we focus on the one above in this book.

9.1 Example Questions

Problem 9.1 There is an urn with 5 green, 6 red, and 4 yellow balls. You pick 3 balls without replacement (that is, without putting the balls back after each pick). Hint: Does this remind you of dealing cards from a deck?

(a) Let A be the event you have at least 3 green balls, and B be the event you have at least 2 green balls. Find $P(A|B)$.

(b) Let A be as above, and B be the event that the first 2 balls picked are green. Find $P(A|B)$.

(c) Compare your answers. How would they compare to $P(A)$? (You do not need to calculate $P(A)$.)

Problem 9.2 Jack is planning to have dinner with Miss Muffet, and will do so as long as he arrives before the spider (who scares her away). Suppose Jack arrives randomly between 5 and 6 PM, and the spider arrives randomly between 5 and 7 PM.

(a) What is the probability Jack has dinner with Miss Muffet?

(b) Given that Jack arrives before 5:30, what is the probability he has dinner with Miss Muffet?

(c) Given that Jack has dinner with Miss Muffet, what is the probability he arrives before 5:30?

Problem 9.3 Suppose an urn contains 2 red, 3 green, and 4 yellow balls. You pick one ball, note it, and replace it (back in the urn). You do this a total of 6 times, what is the probability:

(a) You get exactly 3 green balls.

(b) You get all 6 balls of the same color.

(c) You get the same number of each color.

(d) You get exactly 3 green and 2 yellow balls.

Problem 9.4 Suppose boys and girls are equally likely and each child's gender is independent from the others. Let A be the event that the family has at least one child of each gender. Let B be the event the family has at most one girl. Are A and B independent if:

(a) the family has two children.

(b) the family has three children.

Problem 9.5 Suppose you are dealt two cards from a standard deck. Find and compare the following probabilities:

(a) the probability the first card is a club.

(b) the probability the second card is a club.

(c) the probability the second card is a club, given that the first card is a club.

(d) the probability the first card is a club, given that the second card is a club.

Problem 9.6 Suppose you flip an unfair coin. If you get heads, you roll a fair die twice. If you get tails, you roll a fair die three times.

(a) If you get tails, what is the probability the sum of the rolls is 8?

(b) Suppose the probability of tails is 4 times that of the probability of heads. Find the probability the sum of all the rolls is 8.

Problem 9.7 Suppose that at a given time, 1 in 500 people have strep throat. Further, suppose we have a test for strep throat so that: (i) If the person has strep throat the test works 98% of the time, (ii) If the person does not have strep throat, the test works 96% of the time. Suppose we randomly take someone and test them for strep throat. Let S be the event that a person has strep and $\oplus$ be the event a person tests positive. Thus, S^c is a person not having strep throat and $\oplus^c = \ominus$ is a person testing negative.

(a) Find $P(S), P(\oplus|S), P(\ominus|S)$. Hint: This is a reading comprehension question as well as a math one!

(b) Write out the Bayes' Theorem formula for $P(S|\oplus)$ and calculate this probability.

Problem 9.8 Suppose you roll a fair six-sided die. You then flip a coin the number of times shown on the die.

(a) What is the probability you get exactly 5 heads and 0 tails?

(b) What is the probability you get exactly 5 heads?

(c) Given that you get exactly 5 heads, what is the probability you rolled a 5?

Problem 9.9 Suppose Jack arrives randomly between 5 and 6 PM, the spider arrives randomly between 5 and 7 PM, and Jack gets to have dinner with Miss Muffet as long as he shows up before the spider. Let A be the event that Jack has dinner with Miss Muffet and B be the event that Jack and the spider show up within an hour of each other. Are A and B independent?

Problem 9.10 Suppose Billy is nervous for his first airplane flight. So nervous that he cannot remember his assigned seat when he is the first to board the plane. He randomly sits in one of the 100 seats. Further, every other passenger is very polite, and if Billy (or anyone else) is in their seat will simply sit randomly in one of the remaining.

Suppose you are the last to board the plane. What is the probability you get your assigned seat? Hint: First examine the problem for smaller numbers of seats and try to find a pattern/relationship between them!

9.2 Quick Response Questions

Problem 9.11 Suppose you have a circle of radius 2 within a circle of radius 5. If a dart is thrown randomly into the bigger circle, what is the probability that it lands in the smaller circle? Express your answer as a decimal rounded to the nearest hundredth if necessary.

Problem 9.12 Suppose you are dealt 4 cards from a standard 52 card deck. The probability that all 4 cards are of different rank and have different suits can be expressed as $\frac{P}{Q}$ for positive integers P, Q with $\gcd(P, Q) = 1$. What is $Q - P$?

Problem 9.13 Suppose you have five balls in an urn labeled 1 to 5 and you randomly draw a ball with replacement three times, respecting order. Let the event E denote that the balls drawn add up to 5. Then the probability of E can be written as $\frac{P}{Q}$ for positive integers with $\gcd(P, Q) = 1$. What is $P + Q$?

Problem 9.14 Suppose you role a fair die 3 times. The probability that you get at least one 6 can be expressed as $\frac{P}{Q}$ for positive P, Q with $\gcd(P, Q) = 1$. What is $P + Q$?

Problem 9.15 Which of the following equations can be used to verify independence? That is, which of the following is equivalent to $P(A)P(B) = P(A \cap B)$.

(A) $A \cap B = \emptyset$
(B) $P(A) = P(A|B)$
(C) $P(B) = P(A|B)$ and $P(B) \neq 0$
(D) $P(B) = P(B|A)$ and $P(A) \neq 0$

Problem 9.16 Suppose you role a pair of fair dice. What is the probability that one of the dice comes up 2, given that the sum of the numbers on the dice is 6? Express your answer as $K\%$, where K is rounded to the nearest integer if necessary.

Problem 9.17 Suppose you draw a card from a standard 52 card deck. Let A be the event you draw an ace, and let B_1, B_2, B_3, B_4, respectively, be the event you draw a spade, club, heart, and diamond. Which of the following below is not equal to the others?

(A) $P(B_1)P(A|B_1)+P(B_2)P(A|B_2)+P(B_3)P(A|B_3)+P(B_4)P(A|B_4)$
(B) $P(A|B_1)+P(A|B_2)+P(A|B_3)+P(A|B_4)$
(C) $P(A\cap B_1)+P(A\cap B_2)+P(A\cap B_3)+P(A\cap B_4)$
(D) $P(A)$

Problem 9.18 Suppose you role a fair die once. Let A, B, and C be the events you get an odd number, a number greater than 3, and number greater than 4, respectively. Are A and B independent?

Problem 9.19 Suppose you role a fair die once. Let A and C be the events you get an odd number and a number greater than 4, respectively. Are A and C independent?

Problem 9.20 You have three coins that look identical in an urn. Two of them (type 1) come up heads with probability $1/2$ when you flip them, while the third (type 2) comes up heads with probability $1/4$. Suppose you draw a coin from the urn and flip it, and it comes up heads. Use Bayes' rule to calculate the probability that the coin is of type 2. The probability can be written as $\frac{P}{Q}$ for positive integers P, Q with $\gcd(P,Q)=1$. What is $P+Q$?

9.3 Practice Questions

Problem 9.21 There is an urn with 5 green, 6 red, and 4 yellow balls. You pick 4 balls without replacement (that is, without putting the balls back after each pick). Let A be the event you pick 2 green balls and B be the event you pick 2 yellow balls. Find $P(A|B)$ and $P(B|A)$.

Problem 9.22 Jack and George go to visit Miss Muffet. They both arrive randomly between 5 and 8 PM. Miss Muffet is only available between 6-7 PM. What is the probability that Miss Muffet visits with either Jack or George (or both).

Problem 9.23 Suppose an urn contains 4 red, 3 green, and 2 yellow balls. You pick one ball, note it, and replace it (back in the urn). You do this a total of 3 times, what is the probability:

(a) You get one ball of each color?

(b) You get exactly 2 red balls?

Problem 9.24 Assume boys and girls are equally likely and each child's gender is independent from the others. Let A be the event that the family has at least one child of each gender and C be the event that the family has at most two girls. If a family has four children, which is more likely, $P(A|C)$ or $P(C|A)$?

Problem 9.25 Suppose you are dealt three cards from a standard deck of cards.

(a) Find the probability the third card is a club, given that the first two cards drawn are clubs.

(b) Find the probability the first two cards are clubs, given that the third card drawn is a club.

Problem 9.26 You have an unfair three-sided coin with sides H, T, and E (for heads, tails, and edge). Suppose that the probabilities for the outcome of a toss are $P(H) = 1/2$, $P(T) = 1/3$, and $P(E) = 1/6$. Suppose further that if you get heads, you roll a fair die twice, tails, you roll it three times, and if you get the edge, you roll it four times. Find the probability the sum of the rolls is 8.

Problem 9.27 Suppose that at a given time, 1 in 500 people have strep throat. Further, suppose we have a test for strep throat so that: (i) If the person has strep throat the test works 98% of the time, (ii) If the person does not have strep throat, the test works 96% of the time. Suppose we randomly take someone and test them for strep throat.

Recall the test for strep throat from in class where $P(\oplus|S) = 0.98$ and $P(\oplus|S^c) = .04$. The reason the calculated value of $P(S|\oplus)$ was so low was because we were randomly testing anyone in the population for strep throat. Suppose instead that we only test people for strep throat if they are having symptoms for stream throat, and that 1 in 2 people have strep throat if they are having symptoms (so now assume $P(S) = 1/2$). Calculate $P(S|\oplus)$.

Problem 9.28 Suppose you toss a fair (two-sided) coin. You then roll a fair die once if you get heads and twice if you get tails. What is the probability your coin came up tails if the sum of your rolls is 6?

Problem 9.29 Jack and George go to visit Miss Muffet. They both arrive randomly between 5 and 8 PM. Miss Muffet is only available between 6-7 PM. Let A be the event that Miss Muffet meets only with Jack, and let B be the event that Miss Muffet meets only with George. Are A and B independent?

Problem 9.30 The game odd player out is played as follows: There are n players, each with a coin that has the same probability p of coming up heads on a toss. Everybody tosses their coin at the same time. If there is an "odd player," that is, a player with an outcome different from all of the other players, then that person is eliminated. The remaining players continue until there are two left. At that point, the first player to get heads when the other player gets tails is the winner.

Suppose you are playing odd player out with two other friends and fair coins. What are the chances that you win?

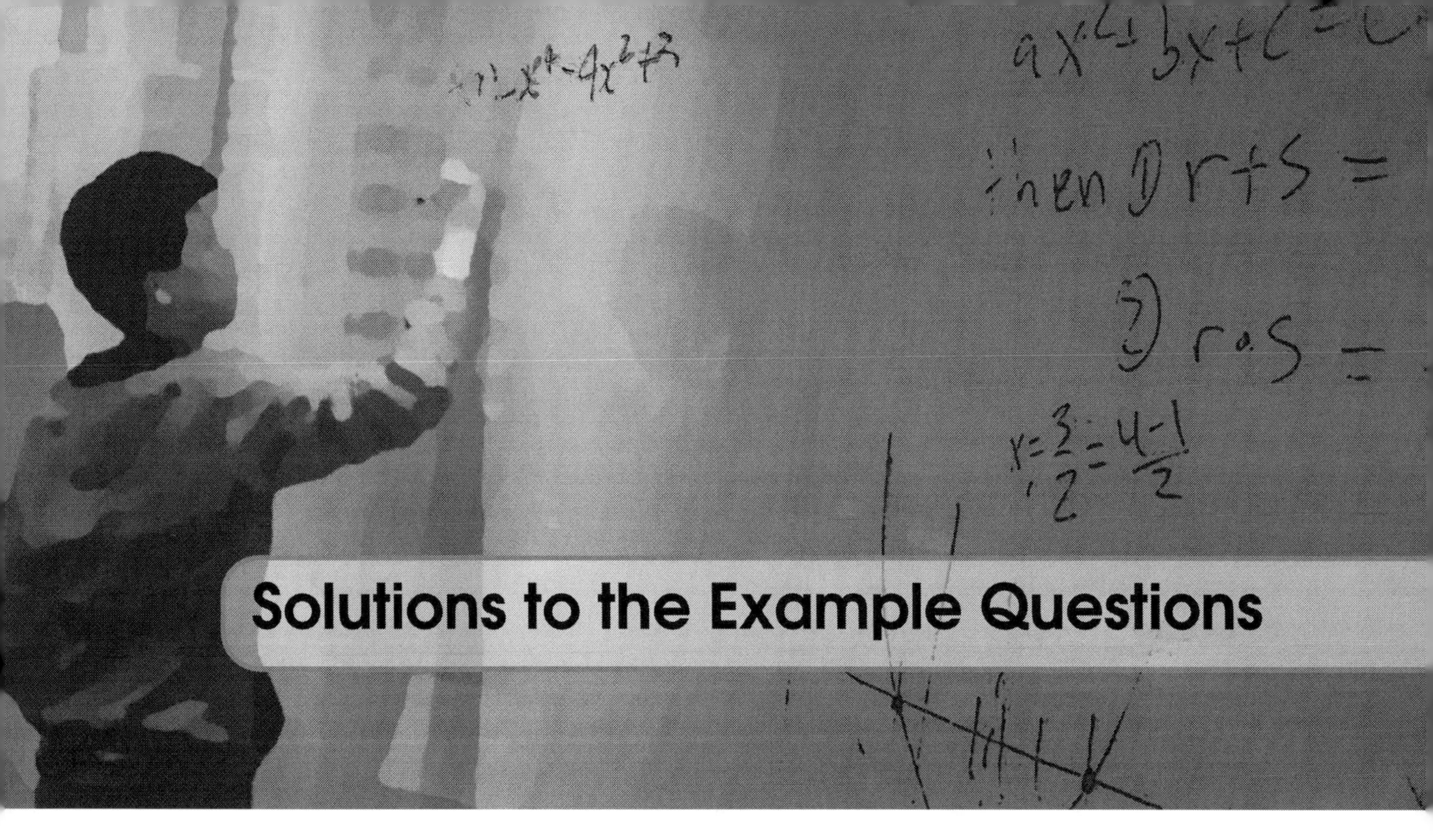

Solutions to the Example Questions

In the sections below you will find solutions to all of the Example Questions contained in this book.

Quick Response and Practice questions are meant to be used for homework, so their answers and solutions are not included. Teachers or math coaches may contact Areteem at info@areteem.org for answer keys and options for purchasing a Teachers' Edition of the course.

1 Solutions to Chapter 1 Examples

Problem 1.1 Suppose John has 2 hats, 4 shirts, 1 jacket, 3 pairs of pants, 5 pairs of shorts, and 4 pairs of shoes.

(a) Suppose John makes an outfit consisting of a shirt, a pair of pants, and a pair of shoes. How many different outfits does he have?

Answer

48.

Solution

We use the product rule: $4 \cdot 3 \cdot 4 = 48$.

(b) Repeat (a) if John *can* wear shorts instead of pants.

Answer

128.

Solution

Using the Sum Rule, we have $3 + 5 = 8$ choices for leg wear. We then proceed as in (a): $4 \cdot 8 \cdot 4 = 128$.

(c) Now suppose John can wear shorts or pants as in (b), *but* if he wears shorts, he will also wear a hat and possibly a jacket.

Answer

368.

Solution

Consider the two cases (so Sum Rule) based on whether John wears shorts or pants. If he wears pants, it is the same as (a); if he wears shorts, we also have to choose which hat he wears and whether or not he wears a jacket (2 choices). The total is: $4 \cdot 3 \cdot 4 + 4 \cdot 5 \cdot 4 \cdot 2 \cdot 2 = 368$.

Problem 1.2 **(a)** Find the number of ways to choose an ordered set of 3 numbers from $\{1,2,3,4\}$.

Answer

24.

Solution

We have $\frac{4!}{1!} = 4 \times 3 \times 2 = 24$. Suppose the numbers are 1, 2 and 3. The possibilities are listed below. Note that it is very convenient to organize a list like this in some order to be certain you haven't left anything out. In this case, they're arranged in a sort of "alphabetical order", as if 1,2,3,4 were A,B,C,D in the alphabet:

$$123, 124, 132, 134, 142, 143, 213, 214, 231, 234, 241, 243,$$
$$312, 314, 321, 324, 341, 342, 412, 413, 421, 423, 431, 432.$$

(b) Find the number of ways to choose 3 numbers from the set $\{1,2,3,4,5,6\}$

Answer

20.

Solution

$\binom{6}{3} = 20$. The possibilities are listed below. Notice that this list is also in "alphabetical order":

$$123, 124, 125, 126, 134, 135, 136, 145, 146, 156,$$
$$234, 235, 236, 245, 246, 256, 345, 346, 356, 456.$$

Problem 1.3 Suppose you have a group of 10 people. How many different photographs are there of everyone lined up if:

(a) all the people look different?

Solution

$10! = 3628800$.

(b) 2 of the people are identical twins who have dressed identically?

Answer

$\frac{10!}{2!} = \binom{10}{2} \cdot 8! = 1814400.$

Solution

Two methods are presented above. For the first, we divide the answer from (a) by 2 because we do not care about the order of the twins (since we cannot tell the twins apart). For the second, we first choose 2 spots for the twins to stand, and then fill in the other 8 people around them.

(c) 2 of the people are a couple and must stand next to each other?

Answer

$2! \cdot 9! = 725760.$

Solution

There are $2 = 2!$ ways to decide the arrangement of the couple. If we then treat them as a single 'object' we must arrange 9 'objects'.

(d) 2 of the people are sworn enemies and cannot stand next to each other?

Answer

$10! - 2! \cdot 9! = 2903040.$

Solution

We count the total number of arrangement and subtract off the number of arrangements where they are next to each other (which is equivalent to them being a couple).

Problem 1.4 How many 2-digit numbers are even? Give at least 2 different proofs.

Answer

45.

Solution

There are $99-10+1=90$ 2-digit numbers and half of them are even. ALTERNATIVELY: There are 9 choices for the first digit $(1,2,\ldots,9)$ and 5 choices for the second digit $(0,2,4,6,8)$ so $9\cdot 5$ total even numbers.

Problem 1.5 How many 10-digit numbers:

(a) do not have the digit 0?

Answer

9^{10}.

Solution

Each digit has 9 choices.

(b) do not have the digit 5?

Answer

$8\cdot 9^9$.

Solution

The first digit cannot be 0 or 5, the rest cannot be 5.

(c) contain the digit 5 exactly 5 times?

Answer

$1\cdot\binom{9}{4}\cdot 9^5+8\cdot\binom{9}{5}\cdot 9^4$.

Solution

We consider cases based on whether the first digit is a 5 or not. We then choose the first digit, which of the remaining digits will be 5, and then choose the remaining digits (which cannot be 5).

(d) contain the digit 5 at least once?

Answer

$9 \cdot 10^9 - 8 \cdot 9^9$.

Solution

There are a total of $9 \cdot 10^9$ 10-digit numbers, and we have already calculated that $8 \cdot 9^9$ of them do not contain 5.

Problem 1.6 How many factors of 2^{20} are larger than $5,000$?

Answer

8.

Solution

Note 2^{20} has 21 factors: $2^0, 2^1, \ldots, 2^{20}$. Further, $2^{12} = 4096 < 5000 < 8192 = 2^{13}$. Hence, 13 of the factors are less than 5000, so $21 - 13 = 8$ are greater.

Problem 1.7 Suppose you write out the numbers $1 - 1000$: $1, 2, 3, 4, \ldots, 1000$.

(a) How many digits have you written in total?

Answer

2893

Solution

$1 \cdot 9 + 2 \cdot 90 + 3 \cdot 900 + 4 \cdot 1 = 2893$.

(b) What is the sum of all the numbers written?

Solution

$$1 + 2 + 3 + \cdot 1000 = \frac{1000 \cdot 1001}{2} = 500500.$$

(c) What is the sum of all the digits written?

Answer

$(0+1+2+3+4+5+6+7+8+9)\cdot 300+1=13501$.

Solution

1000 contributes a digit sum of 1. We may think of all the other numbers as containing 3 digits (including leading 0's if needed) and we can also include 000 without changing the sum. Each digit is thus written a total of $3\cdot 10^2=300$ times.

Problem 1.8 Suppose you have a 5-letter word.

(a) How many possible words are there?

Solution

26^5.

(b) How many of the words do not have consecutive consonants or consecutive vowels (the vowels are a, e, i, o, u)? (That is, suppose the letters alternate between consonant and vowel.)

Answer

$21\cdot 5\cdot 21\cdot 5\cdot 21+5\cdot 21\cdot 5\cdot 21\cdot 5=21^3\cdot 5^2+21^2\cdot 5^3$.

Solution

Consider cases based on whether the word starts with a consonant or a vowel.

(c) How many words have exactly two vowels?

Answer

$\binom{5}{2}\cdot 21^3\cdot 5^2$.

Solution

First, choose where the vowels appear in the word, then pick the 3 consonants and 2 vowels.

(d) How many words have a block of three consecutive consonants and two consecutive vowels.

Answer

$2 \cdot (21^3 \cdot 5^2) = 463050$.

Solution

Note the word will either be cccvv or vvccc (2 ways). In both cases there are $21^3 \cdot 5^2$ ways to pick the letters.

Problem 1.9 Suppose you have a standard 8×8 chessboard.

(a) How many ways can you place 8 mutually non-attacking rooks on the chessboard? (Consider the rooks to be identical.)

Answer

$8! = 40320$.

Solution

Consider one row at a time.

(b) How many ways can you place 2 mutually non-attacking rooks on the chessboard?

Answer

$64 \cdot 49/2! = 1568$.

Solution

The first rook can be on any of the 64 squares, and this rook removes $8+8-1 = 15$ squares for the second rook. We divide by 2! as the rooks are identical.

(c) How many ways can you place 7 mutually non-attacking rooks on the chessboard? (Consider the rooks to be identical.)

Answer

$8 \cdot 8! = 322560$.

Solution

There are 8! ways of placing 8 mutually non-attacking rooks on the board, and then 8 choices for which rook to remove so that 8 remain. Note: This works because for any 7 rooks on the chessboard there is only one place you could put an 8th rook.

Problem 1.10 Telephone numbers in the Land of Nosix have 7 digits, and the only digits available are $\{0,1,2,3,4,5,7,8\}$. No telephone number may begin in $0,1$, or 5. Find the number of telephone numbers possible that meet the following criteria:

(a) you may have repeated digits.

Solution

$5\cdot 8^6 = 1310720.$

(b) you may not have repeated digits.

Solution

$5\cdot 7\cdot 6\cdot 5\cdot 4\cdot 3\cdot 2 = 5\cdot \frac{7!}{1!} = 25200.$

(c) you may have repeated digits, but the phone number must be even.

Solution

$5\cdot 8^5\cdot 4 = 655360.$

(d) you may not have repeated digits, and the phone number must be odd.

Answer

$5\cdot 6\cdot 5\cdot 4\cdot 3\cdot 2\cdot 2+4\cdot 6\cdot 5\cdot 4\cdot 3\cdot 2\cdot 2 = 5\cdot \frac{6!}{1!}\cdot 2+4\cdot \frac{6!}{1!}\cdot 2 = 12960.$

Solution

We now pick the *last* digit first, which must be chosen from $1,3,5,7,9$. We deal with two cases: (i) the last digit is 1 or 5, and (ii) otherwise.

2 Solutions to Chapter 2 Examples

Problem 2.1 There are five Golden retrievers, six Irish setters, and eight Poodles at the pound.

(a) How many total ways are there to choose 2 dogs (in no particular order)?

Solution

$\binom{19}{2} = 171.$

(b) How many ways can two dogs be chosen if they are not the same kind?

Answer

$5 \cdot 6 + 5 \cdot 8 + 6 \cdot 8 = 118.$

Solution

Consider 3 cases based on which two breeds of dog are chosen.

(c) How many ways can two dogs be chosen so that not both of them are Poodles.

Answer

$\binom{19}{2} - \binom{8}{2} = 171 - 28 = 143.$

Solution

There are $\binom{19}{2}$ total ways to choose two dogs. There are $\binom{8}{2}$ ways to choose two Poodles. We subtract to get the number of ways *without* two Poodles.

Problem 2.2 Suppose you have a student group with 15 males and 10 females.

(a) How many ways are there to pick a group of 5 males and 5 females?

Answer

$\binom{15}{5} \cdot \binom{10}{5} = 756756.$

Solution

Pick the groups of males and females separately.

(b) How many ways are there to pick an Executive Committee of 5 members and a Party Planning Committee of 5 members? Members can be on both committees at once, but each committee must have at least one male and at least one female.

Answer

$$\left(\binom{25}{5}-\binom{15}{5}-\binom{10}{5}\right)^2 = 2487515625.$$

Solution

First note there are the same number of ways to choose either committee. We use complementary counting to subtract the number of committees made of only males or only females.

(c) Suppose you still need to pick an Executive Committee and Party Planning Committee (each with 5 members). This time, only the Executive Committee is required to have a member of each gender, but now members are *not* allowed to be on both committees at once.

Answer

$$\left(\binom{25}{5}-\binom{15}{5}-\binom{10}{5}\right)\cdot\binom{20}{5} = 773262000.$$

Solution

Picking the Executive Committee is the same as in part (b), and now the Party Planning Committee is chosen from the remaining 20 members.

Problem 2.3 Suppose you have a group of 8 people. How many different photographs are there of everyone lined up if:

(a) 2 of the people are identical twins and 3 of the people are identical triplets (the twins and triplets dress identically)?

Solution

$$\frac{8!}{3!\cdot 2!} = \binom{8}{3}\cdot\binom{5}{2}\cdot 3! = 3360.$$

(b) the 8 people are 3 singles, a couple and the last 3 of the people are a family (2 parents and a child). The couple must be together and the family must stand together, with the child in between the parents?

Answer

$2! \cdot 2! \cdot 5!$.

Solution

There are only 2! ways to decide the arrangement of the couple and the family (since the child is in the middle). We then rearrange 5 'objects' $(3+1+1)$.

Problem 2.4 5 boys and 3 girls run in a race.

(a) If all the boys finish as a group, how many outcomes are there?

Answer

$4! \cdot 5!$

Solution

There are 4! ways the girls and the one group of boys can be ordered. Then there are 5! different orderings with the boys.

(b) If none of the girls finish right after another (2nd and 3rd, 5th and 6th, etc.), how many outcomes are there?

Answer

$5! \cdot 6 \cdot 5 \cdot 4$.

Solution

First order the 5 boys (5! ways). This creates 6 spaces where the girls could finish, and there are $6 \cdot 5 \cdot 4$ ways to fill in these spaces.

Problem 2.5 The following illustration is a map of a city, and you would like to travel from the southwest (lower left) to the northwest (upper right) of the city along the roads in the shortest possible distance.

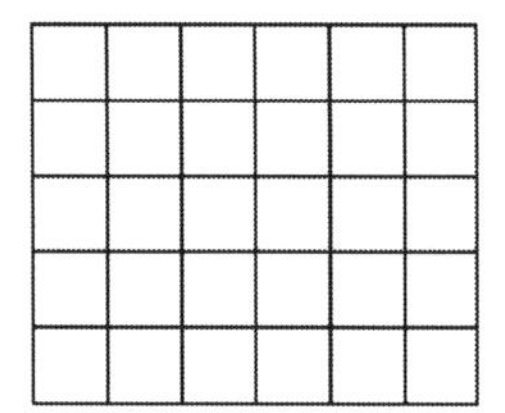

(a) In how many ways can you do this?

Solution

$\binom{11}{6} = 462$. Since we want the shortest possible distance, the path is always moving right or up. There are 11 total "blocks" or "steps" along the way, and 6 of those steps must be right.

(b) Suppose you want to avoid the roads furthest to the west and north. How many different paths remain?

Answer

$\binom{9}{5} = 126$.

Solution

To avoid the roads, the first step must be right, and the last step must be up. Of the 9 remaining steps, 5 must be right.

(c) How many rectangles are in the diagram?

Answer

$\binom{6}{2} \cdot \binom{7}{2} = 315$.

Solution

Each rectangle is determined by two horizontal lines and two vertical lines.

Problem 2.6 Suppose in a group of soldiers you have 3 officers, 6 sergeants, and 30 privates.

(a) How many ways can a team be formed consisting of 1 officer, 4 sergeants, and 10 privates?

Solution

$$\binom{3}{1}\binom{6}{4}\binom{30}{10} = 1352025675.$$

(b) Repeat (a) if the 6 sergeants each lead 5 different privates (with no overlap so the total number of privates is still 30) and privates can only be chosen if their lead sergeant is on the team as well.

Answer

$$\binom{3}{1}\binom{6}{4}\binom{20}{10} = 8314020.$$

Solution

We proceed identically to before except now we only have a pool of 20 privates to choose from.

Problem 2.7 Suppose you have 15 distinct playing cards. You want to divide them into 3 (unordered) groups. How many ways to do this are there if:

(a) the groups are of size 6, 5, and 4?

Answer

$$\binom{15}{6}\binom{9}{5}\binom{4}{4}$$

Solution

Simply pick 6 for the group of size 6, then 5 of the remaining cards for the group of size 5, and the last 4 cards make up the group of size 4.

(b) all three groups have size 5?

Answer

$$\binom{15}{5}\binom{10}{5}\binom{5}{5} \div 3!$$

Solution

Start using an identical method to part (a). Then note that since all three groups are the same size, we must divide by 3! as they are in no particular order.

Problem 2.8 A bookshelf contains 4 German books, 6 Spanish books, and 7 French books. Each book is different from one another.

(a) How many different arrangements can be done of these books?

Solution

$17!$.

(b) How many different arrangements can be done if books of each language must be next to each other?

Answer

$(4!\cdot 6!\cdot 7!)\cdot 3!$.

Solution

First order each language of books, then group them as a single 'object'.

(c) How many different arrangements can be done if no two German books must be next to each other?

Answer

$13!\cdot(14\cdot 13\cdot 12\cdot 11)$.

Solution

First arrange the 13 Spanish and French books. Putting these books creates $13+1=14$ 'spaces' (before the first book, after the last book, and in between the others). Since the German books cannot be next to each other, the four German books must be put in these spaces (at most one book per space).

Problem 2.9 Suppose a school offers the following classes: English, Latin, Algebra, Geometry, Calculus, History, Art, and Music. A student must take a total of 5 classes.

(a) If the order that the classes are chosen doesn't matter and at least one course must be a math course, how many different choices for classes does the student have?

Answer

$\binom{8}{5} - \binom{5}{5} = 55.$

Solution

Subtract the $\binom{5}{5} = 1$ way that a student does not take a math class from the number of choices with no restrictions.

(b) Suppose now the order of the classes does matter and that the student must take exactly one math class. How many choices for a schedule does the student have?

Answer

$\binom{3}{1} \cdot \binom{5}{4} \cdot 5! = 1800.$

Solution

First choose which classes the student will take, and then arrange their order.

Problem 2.10 How many ways are there to arrange the numbers $21, 31, 41, 51, 61$ so that the sum of each consecutive group of 3 numbers is divisible by 3?

Answer

$2 \cdot (2! \cdot 2!) = 8.$

Solution

Mod 3, our numbers are $0, 1, 2, 0, 1$. Arranging the numbers mod 3, the only two possibilities are $1, 0, 2, 1, 0$ or $0, 1, 2, 0, 1$ (convince yourself of this, starting with why the middle number must be 2). From here, we arrange the "0's" $(21, 51)$ and the "1's" $(31, 61)$.

3 Solutions to Chapter 3 Examples

Problem 3.1 Suppose 8 dinner guests attend a dinner party and are seated at a circular table. How many ways are there to seat the guests if:

(a) none of the seats are special, we only care about how they are arranged among themselves?

Answer

$\frac{8!}{8} = 5040.$

Solution

This is a circular permutation.

(b) one of the seats is the "Head of the Table"?

Answer

$8! = 8 \cdot \frac{8!}{8} = 40320.$

Solution

Note, once we distinguish one seat, we can tell them all apart. Alternatively, if we first arrange the 8 guests in a circle, there are 8 choices for rotating the circle to choose who gets the special seat.

(c) none of the seats are special, but two of the guests are a couple and want to sit next to each other?

Answer

$2! \cdot \frac{7!}{7} = 1440.$

Solution

There are 2! ways to arrange the couple, and then we treat them as a single object.

Problem 3.2 The number 3 can be expressed as a sum of one or more positive integers in four ways, namely, as 3, $1+2$, $2+1$, and $1+1+1$.

(a) How many ways can the number 5 be expressed as the sum of one or more positive integers less than or equal to 2?

Answer

$\binom{5}{0}+\binom{4}{1}+\binom{3}{2}=1+4+3=8.$

Solution

The integers must all be 1 or 2. We break into 3 cases based on how many 2's are used (either $0,1,2$). In these cases (respectively) we know there are a $5,4,3$ total integers used, so we pick which of those integers will be 2.

(b) How many total ways can 5 be expressed as the sum of one or more positive integers?

Answer

16.

Solution

Note that there are many more cases to deal with here (try to work them out yourself!). We use a trick: $5=1+1+1+1+1$, and we can consider the "1"s as 'sticks", and if several "1"s are together we treat them as the number of sticks in the same pile. At each gap between two "1"s, the plus sign can be either present or absent, so there are 2 choices for each gap, and there are 4 gaps. Therefore there are $2^4=16$ possible sums.

(c) How many total ways can 5 be expressed as the sum of two non-negative integers?

Answer

6

Solution

We can write

$$5=0+5=1+4=2+3=3+2=4+1=5+0.$$

Note we can also use stars and bars.

Problem 3.3 Consider the number 6.

(a) How many ways can you express 6 as the sum of three non-negative integers?

Answer

28

Solution

Using stars and bars the answer is

$$\binom{6+3-1}{6} = \binom{6+3-1}{3-1} = \binom{8}{6} = \binom{8}{2} = 28.$$

(b) How many ways can you express 6 as the sum of three positive integers?

Answer

28

Solution

Using stars and bars the answer is

$$\binom{6-1}{3-1} = \binom{5}{2} = 10.$$

Problem 3.4 Beagel likes bagels, and he went to the Bagel Shop to buy 6 bagels for breakfast. The Bagel Shop sells 3 types of bagels: sourdough, blueberry, and sesame seeds.

(a) If Beagel plans to buy at least one of each type. In how many ways can he do this?

Answer

10.

Solution

$a+b+c=6$ and a,b,c are all positive integers. This is a stars and bars problem, and the answer is $\binom{5}{2} = 10$. The answer can also be obtained by listing all the possibilities.

(b) If Beagel does not need to buy at least one of each type, how many ways can he buy the 6 bagels?

Answer

$\binom{6+3-1}{6} = 28.$

Solution

Now we have $a+b+c=6$ were a,b,c are all non-negative integers. It is still stars and bars, but the alternate version.

(c) How many ways are there to buy the bagels so that Beagel gets at least 2 types?

Answer

$\binom{6+3-1}{6} - 3 = 25.$

Solution

There are only 3 of the total outcomes where Beagel does not have at least 2 types.

Problem 3.5 Suppose you have 30 identical balls and 6 numbered boxes. How many ways are there to put the balls into the boxes if:

(a) there are no restrictions?

Answer

$\binom{30+6-1}{30} = 324632.$

Solution

This is routine stars and bars.

(b) each box has at least two balls?

Answer

$\binom{18+6-1}{18} = 33649.$

Solution

First put two balls in each box. We are left with 18 balls and 6 boxes.

(c) no box has more than 5 balls?

Answer

1.

Solution

Note the only possibilities here is that each box has exactly 5 balls.

(d) the first box has exactly 10 balls?

Answer

$\binom{20+5-1}{20} = 10626.$

Solution

First put 10 balls into the first box. We are left with 20 balls and 5 boxes.

Problem 3.6 Consider the number 10000.

(a) How many factors does it have?

Answer

25.

Solution

$10000 = 2^4 \cdot 5^4$ so it has $(4+1)\cdot(4+1) = 25$ factors.

(b) How many ways are there to represent it as the product of 2 factors if we consider products that differ in the order of factors to be different?

Answer

25.

Solution

Note this is the same as part (a) as each factor comes in a pair. ALTERNATIVELY: $10000 = 2^4 \cdot 5^4 = a \cdot b = (2^m 5^n)(2^p 5^q)$, so it is actually stars and bars for both the powers of 2 and powers of 5: $m + p = 4$ and $n + q = 4$, zeros are allowed. Therefore the answer is $\binom{4+2-1}{4}^2 = 25$.

(c) How many ways are there to represent it as the product of 3 factors if we consider products that differ in the order of factors to be different?

Answer

225.

Solution

$10000 = 2^4 \cdot 5^4 = a \cdot b \cdot c = (2^m 5^n)(2^p 5^q)(2^r 5^s)$, so we need $m + p + r = 4$ and $n + q + s = 4$, zeros are allowed. Using stars and bars, the answer is $\binom{4+3-1}{4}^2 = 225$.

Problem 3.7 Suppose you have 5 blue, 5 red, and 5 green balls. You want to arrange the balls so that no two green balls are next to each other. How many ways are there to do this if

(a) the balls are in a row and each ball is numbered?

Answer

$10! \cdot \dfrac{11!}{6!} = 201180672000$.

Solution

Arrange the blue and red balls first. This creates $10 + 1$ spaces for the green balls. We then arrange the green balls in those spaces.

(b) the balls are in a row and each ball is identical?

Answer

$\binom{10}{5} \cdot \binom{11}{5} = 116424,$

Solution

Same strategy as part (a), except since the balls are identical we choose where to place the red and green balls. ALTERNATIVELY: We divide the answer in part (a) by $(5!)^3$, as we do not care about how the balls are numbered. Double check this leads to the same answer!

(c) the balls are in a circle and each ball is numbered?

Answer

$\frac{10!}{10} \cdot \frac{10!}{5!} = 10973491200.$

Solution

Arrange the blue and red balls first (this is a circular permutation). Since they are arranged in a circle, this creates 10 spaces for the green balls.

Problem 3.8 Suppose you have 30 numbered balls and 6 numbered boxes. How many ways are there to put the balls into the boxes if:

(a) there are no restrictions?

Answer

$6^{30} = 221073919720733357899776.$

Solution

There are 6 choices for each ball.

(b) no box has more than 5 balls?

Answer

$\frac{30!}{(5!)^6} = \binom{30}{5} \cdot \binom{25}{5} \cdots \binom{5}{5} = 88832646059788350720.$

Solution

Note each box must have exactly 5 balls. Therefore, we need to divide the balls into 6 groups of 5.

Problem 3.9 Suppose 5 girls and 15 boys sit around the table. How many arrangements are there

(a) in total?

Answer

$\frac{20!}{20} = 121645100408832000.$

Solution

This is a circular permutation.

(b) if there is at least 1 boy in between all the girls?

Answer

$\frac{5!}{5} \cdot \binom{14}{4} \cdot 15! = 31415569016832000.$

Solution

First arrange the girls as a circular permutation. Then choose where the boys will sit (but not arranging the boys yet): the girls create 5 spaces and we place 15 boys in those spaces, ensuring each space has at least one boy (using stars and bars). Lastly, we arrange the 15 boys.

Problem 3.10 Suppose 10 people get in an elevator on Floor 0. The people leave the elevator somewhere between (inclusive) Floors 1 and Floor 5.

(a) If we only care about how many people get of at each floor, how many ways can the people get off?

Answer

$\binom{10+5-1}{10} = 1001.$

Solution

This is stars and bars.

(b) If we only care about what collection of floors the elevator stops on, how many different collections are there?

Answer

$2^5 - 1 = 31$.

Solution

Note that we have enough people that it is possible for every floor to be stopped on. Hence, it is possible for the elevator to stop or not (2 choices) on each floor. However, it must stop on at least one floor, so we subtract 1.

4 Solutions to Chapter 4 Examples

Problem 4.1 Suppose you have 8 pairs of people (so 16 people in total). How many ways are there to arrange the pairs (each pair next to each other) in a circle

(a) in total?

Answer

$(2!)^8 \cdot \frac{8!}{8}$.

Solution

There are 2! ways to arrange each pair, and then arrange the 4 pairs in a circle.

(b) if each pair is a different set of identical twins (that dress alike)?

Answer

$\frac{8!}{8}$.

Solution

This is the same as (a) except all the pairs only have 1 way to be arranged.

(c) Repeat (a) if there is a special "Top" position in the circle that is different from the other places.

Answer

$(2!)^8 \cdot \frac{8!}{8} \cdot 16$.

Solution

After (a), we simply choose which of the 8 people is in the top position.

(d) Repeat (b) if there is a special "Top" position in the circle that is different from the other places.

Answer

$\frac{8!}{8} \cdot 16$.

Solution

After (b), we simply choose which of the 8 people is in the top position.

Problem 4.2 **(a)** Work out and write out the PIE formula for 3 sets A, B, C.

Solution

$n(A \cup B \cup C) = n(A) + n(B) + n(C) - n(A \cap B) - n(A \cap C) - n(B \cap C) + n(A \cap B \cap C)$.

(b) How many terms will the PIE formula for 4 sets A, B, C, D have?

Answer

15

Solution

There are $\binom{4}{1} = 4$ terms involving 1 set, $\binom{4}{2} = 6$ terms involving 2 sets, $\binom{4}{3} = 4$ involving 3 sets, and $\binom{4}{4} = 1$ with all 4 sets. Hence there are a total of $4+6+4+1 = 15$ terms.

Problem 4.3 Using the PIE formula, find the number of positive integers between 1 and 1000 that are either a multiple of 5, a multiple of 6, or a multiple of 7.

Answer

$200 + 166 + 142 - 33 - 28 - 23 + 4 = 428$.

Solution

Let A be the multiples of 5, B the multiples of 6, and C the multiples of 7. We then have $n(A) = 200, n(B) = 166, n(C) = 142$. Note that $A \cap B$ is all the multiples of 30, $A \cap C$ multiples of 35, $B \cap C$ multiples of 42, so $n(A \cap B) = 33, n(A \cap C) = 28, n(B \cap C) = 23$. Lastly, $A \cap B \cap C$ is the multiples of 210, so $n(A \cap B \cap C) = 4$. Hence, $n(A \cup B \cup C) = 200 + 166 + 142 - 33 - 28 - 23 + 4 = 428$.

Problem 4.4 Suppose Albert, Bill, and Charles run a race.

(a) How many different outcomes of the race are there?

Answer

$3! = 6$.

Solution

This is a permutation.

(b) Suppose now they run a second race. How many different outcomes for the 2nd race are there if no one finished the second race in the same place as the first race?

Answer

2.

Solution

It is possible to list all the outcomes. We can also use PIE to calculate the number of ways so that at least one person finishes in the same place, and subtract it from the $3! = 6$ total outcomes. Let A, B, C be be the event Albert, Bill, Charles (respectively) finish in the same position as the first race. Then, using PIE we have

$$n(A \cup B \cup C) = \binom{3}{1} \cdot 2! - \binom{3}{2} \cdot 1! + \binom{3}{3} = 3! - \frac{3!}{2!} + \frac{3!}{3!} = 4.$$

Thus the final answer is $6 - 4 = 2$.

Problem 4.5 How many ways are there to put numbered 6 balls in 3 numbered boxes, so that each box gets at least one ball? Hint: It is probably easiest to use PIE and Complementary Counting.

Answer

$3^6 - 3 \cdot 2^6 + 3 = 540$.

Solution

Let A, B, C be respectively events that the three boxes get 0 balls. Then $n(A) = n(B) = n(C) = 2^6$, and $n(A \cap B) = n(A \cap C) = n(B \cap C) = 1^6$ and $n(A \cap B \cap C) = 0$. Hence $n(A \cup B \cup C) = 3 \cdot 2^6 - 3$. We then subtract this from the total number of outcomes.

Problem 4.6 Suppose you have a set $S = \{1, 2, 3, \ldots, 40\}$. You want to choose A, B, C

such that $A \cup B \cup C = S$ and $A \cap B \cap C = \emptyset$. (Remember $\emptyset = \{\}$ is the empty set, which contains no elements.) How many ways an this be done

(a) if we also assume $A \cap B = A \cap C = B \cap C = \emptyset$? (Under these conditions, A, B, C *partitions* S.)

Answer

3^{40}.

Solution

Every number $1, 2, 3, \ldots, 40$ is either in A or B or C.

(b) in total?

Answer

6^{40}.

Solution

Think of the Venn diagram. Every number $1, 2, 3, \ldots, 40$ can either be in just A, just B, just C, or in $A \cap B$, in $A \cap C$, or in $B \cap C$ (6 choices for each).

Problem 4.7 Suppose you have the numbers $\{1, 2, 3, 4, 5, 6\}$.

(a) How many 6-digit numbers can be formed using each number once?

Answer

$6! = 720$.

Solution

This is a permutation.

(b) How many 6-digit numbers can be formed with 2 next to 1 or 3? (Again use each number above once.) Hint: PIE.

Answer

$2! \cdot 5! + 2! \cdot 5! - 2! \cdot 4! = 432$.

Solution

Let A, B be 2 is adjacent to $1, 3$ respectively. Then $n(A) = n(B) = 2! \cdot 5!$ if we think of $2, 1$ or $3, 1$ as a pair. For $A \cap B$, group $1, 2, 3$ with 2 in the middle, so $n(A \cap B) = 2! \cdot 4!$.

(c) How many 6-digit numbers can be formed with 2 not next to 1 or 3? (Again use each number above once.)

Answer

$6! - 2! \cdot 5! - 2! \cdot 5! + 2! \cdot 4! = 288$.

Solution

Combine parts (a) and (b) and use complementary counting.

Problem 4.8 Suppose that students take three tests in a course and that exactly 11 students get A's on each exam. How many students must get A's on all three exams if exactly 9 students get A's on any two exams and 14 students get an A on at least one exam?

Answer

8

Solution

Let X, Y, Z, be, respectively, the set of students that get an A on the first exam, the second exam, and the third exam. Then $n(X \cap Y \cap Z) = n(X \cup Y \cup Z) - n(X) - n(Y) - n(Z) + n(X \cap Y) + n(X \cap Z) + n(Y \cap Z) = 14 - 3 \cdot 11 + 3 \cdot 9 = 8$.

Problem 4.9 Suppose some friends go to a party. They each wear a coat. However, as they are leaving, they each randomly grab a coat. How many ways can the friends leave so that *none* of them have their own coat, if there are:

(a) 3 friends?

Answer

2.

Solution

It is possible to list all the outcomes. We can also use PIE to calculate the number of ways for them to leave so that at least one person has their own coat, and subtract it from the $3! = 6$ total outcomes. Let A, B, C be be the event the 1st, 2nd, and 3rd person (respectively) gets their own coat back. Then, using PIE we have

$$n(A \cup B \cup C) = \binom{3}{1} \cdot 2! - \binom{3}{2} \cdot 1! + \binom{3}{3} = \frac{3!}{1!} - \frac{3!}{2!} + \frac{3!}{3!} = 4.$$

Thus the final answer is

$$2 = 6 - 4 = \frac{3!}{0!} - \frac{3!}{1!} + \frac{3!}{2!} - \frac{3!}{3!}$$

(b) 4 friends?

Answer

9.

Solution

Using a similar method to that in part (a), (using A, B, C, D for 1st, 2nd, 3rd, 4th person), we have

$$n(A \cup B \cup C \cup D) = \binom{4}{1} \cdot 3! - \binom{4}{2} \cdot 2! + \binom{4}{3} \cdot 1! + \binom{4}{4} = \frac{4!}{1!} - \frac{4!}{2!} + \frac{4!}{3!} - \frac{4!}{4!} = 15.$$

Thus the final answer is

$$9 = 24 - 15 = \frac{4!}{0!} - \frac{4!}{1!} + \frac{4!}{2!} - \frac{4!}{3!} + \frac{4!}{4!}$$

Problem 4.10 Suppose you have copies of the 7 Harry Potter books. You give out the books to 3 of your friends. Each friend gets at least one book. How many ways can you give out the books? (The books you give each friend are not in any specific order.)

Answer

$3^7 - 3 \cdot 2^7 + 3 = 1806.$

Solution

Let A,B,C be respectively events that the three friends get 0 books. Then $n(A)=n(B)=n(C)=2^6$, and $n(A\cap B)=n(A\cap C)=n(B\cap C)=1^6$ and $n(A\cap B\cap C)=0$. Hence $n(A\cup B\cup C)=3\cdot 2^6-3$. We then subtract this from the total number of outcomes (3^7).

5 Solutions to Chapter 5 Examples

Problem 5.1 A train with 20 passengers must make 7 stops.

(a) How many ways are there for the passengers to get off the train at the stops?

Answer

$7^{20} = 79792266297612001$.

Solution

Each passenger has 7 choices for when to get off.

(b) Repeat part (a) if we only care about the number of passengers getting off at each stop?

Answer

$\binom{20+7-1}{20} = 230230$.

Solution

Now the passengers are considered identical so we use stars and bars.

Problem 5.2 Suppose you have four black, four white, and four green balls. Assume balls of the same color are identical. How many ways are there to put all 12 balls into the 6 distinguishable boxes if

(a) if all 12 balls were black (that is, all identical)?

Answer

$\binom{12+6-1}{12} = 6188$.

Solution

This is stars and bars.

(b) if each box can have at most one ball of the same color? (Color matters!)

Answer

$$\binom{6}{4}^3 = 3375.$$

Solution

For each color, choose which 4 boxes the balls go in.

(c) if multiple balls of the same color can be in the same box?

Answer

$$\binom{4+6-1}{4}^3 = 2000376.$$

Solution

We use stars and bars for each color separately.

Problem 5.3 The following problems are about a very smart grasshopper.

(a) Suppose a grasshopper sits at the center O of one corner of an 8×8 chessboard. At a given moment, it can jump to the center of any of the squares which have a common edge with the square where it currently sits, as long as the jump increases the distance between point O and the position of the grasshopper. How many ways are there for the grasshopper to reach the square at the opposite corner?

Answer

$$\binom{14}{7} = \frac{14!}{(7!)^2} = 3432.$$

Solution

From one corner to the other it is 14 steps (7 vertically and 7 horizontally).

(b) A $8 \times 8 \times 8$ cube is formed of small unit cubes. A grasshopper "sits" at the center O of one corner cube. At a given moment, it can "jump" to the center of any of the cubes which have a common face with the cube where it currently sits, as long as the jump increases the distance between point O and the position of the grasshopper. How many ways are there for the grasshopper to reach the cube at the opposite corner?

Answer

$\binom{21}{7}\cdot\binom{14}{7}\cdot\binom{7}{7}=\frac{21!}{(7!)^3}=399072960.$

Solution

From one corner to the other it is 21 steps (7 each in the x, y, z directions).

Problem 5.4 Suppose you have 6 red cards and 20 black cards. Assume all the cards of the same color are identical. Deal the cards out in a line.

(a) How many different arrangements of the cards are there?

Answer

$\binom{26}{6}=230230.$

Solution

Since cards of the same color are identical, decide which 6 of the 26 slots are for the red cards.

(b) Repeat part (a), if there must be at least 2 black cards between all the red cards.

Answer

$\binom{10+7-1}{10}=8008.$

Solution

Arrange the red cards (only 1 way). Then place 2 black cards in between the cards (only 1 way) using $2\cdot 5=10$ black cards. Then the remaining 10 cards can arranged in the 7 spaces created by the red cards (now including the endpoints) using stars and bars.

Problem 5.5 Given positive integers $1,2,3,\ldots,10$. Let a permutation of these numbers satisfy the requirement that, for each number, it is either (i) greater than all the numbers after it, or (ii) less than all the numbers after it. How many such permutations are there? For example, $1,2,3,4,5,6,7,8,9,10$ and $10,1,2,3,4,5,6,7,8,9$ both work.

Answer

512

Solution

Start with the first number: it is either 1 or 10 (2 choices). For the second number, it is either the largest remaining, or the smallest remaining (2 choices again). This pattern continues for all the numbers except the 10th. There are thus $2^9 = 512$ permutations.

Problem 5.6 Suppose you have 10 identical balls. How many ways are there to put them in 7 numbered boxes so that at least one of the boxes gets at least 4 balls?

Answer

5880

Solution

Let $A, B, \ldots, G$ be the events that box $1, 2, 3, \ldots 7$ (respectively) get at least 4 balls. We want $n(A \cup B \cup \cdots \cup G)$. Note

$$n(A) = \cdots = n(G) = \binom{6+7-1}{6}$$

using stars and bars (put 4 balls in the respective box, and then arrange the remaining $10 - 4 = 6$ balls in any of the boxes). Similarly we have

$$n(A \cap B) = \cdots = n(F \cap G) = \binom{2+7-1}{2}.$$

Since the intersection of 3 or more of these sets is empty,

$$n(A \cup B \cup \cdots \cup G) = \binom{7}{1} \cdot \binom{12}{6} - \binom{7}{2} \cdot \binom{8}{2} = 5880,$$

as our final answer.

Problem 5.7 Recall Problem 5.4 where we had 6 red cards and 20 black cards. Now suppose the 6 red cards are numbered and the cards are dealt out in a circle. (The black cards are still identical.)

(a) How many different arrangements of the cards are there?

Answer

$\frac{6!}{6} \cdot \binom{20+6-1}{20} = 6375600.$

Solution

Arrange the 6 red cards in a circle ($6!/6$ ways) creating 6 spaces. The arrange the black cards in those spaces using stars and bars.

(b) Repeat part (a), if there must be at least 2 black cards between all the red cards.

Answer

$\frac{6!}{6} \cdot \binom{8+6-1}{8} = 154440.$

Solution

Arrange the red cards (still $6!/6$ ways). Then place 2 black cards in between the cards (only 1 way) using $2 \cdot 6 = 12$ black cards. Then the remaining 8 cards can arranged in the 6 spaces created by the red cards (now including the endpoints) using stars and bars.

Problem 5.8 How many ways are there to write 201 as the sum of three non-negative integers (we care about the order of the numbers) if all three numbers must be different?

Answer

$\binom{201+3-1}{201} - 300 - 1 = 20202.$

Solution

There is 1 way ($201 = 67+67+67$) for 201 to be written as the sum of one repeated number. Now consider having a pair of repeated numbers. That is we have $a+a+b = 2a+b = 201$ for non-negative a,b. We can list the outcomes as ordered pairs (a,b): $(0,201),(1,199),(2,197),\dots(67,67),\dots,(100,1)$, that is, there are $101-1$ new possibilities (as we already dealt with $201 = 67+67+67$ above). We subtract this from the total number of outcomes in (a). Each of these can be arranged in 3 different ways, for a total of 300 ordered outcomes. We subtract these two cases from the total number of outcomes.

Problem 5.9 Consider the number 16000000.

(a) How many ways are there to represent it as the product of two factors, each divisible by 8, if we consider products that differ in the order of factors to be different?

Answer

$\binom{4+2-1}{4} \cdot \binom{6+2-1}{6} = 35.$

Solution

$16000000 = 2^{10} \cdot 5^6 = (2^m \cdot 5^n)(2^p \cdot 5^q)$. Thus we need $m+p = 10, n+q = 6$ with $m, p \geq 3$ and $n, q \geq 0$. Hence use stars and bars.

(b) Repeat (a) if we do *not* care about the order of the factors.

Answer

18.

Solution

Note that the number is a perfect square. All other pairs of factors can be reversed, so the total is $1+(35-1)/2$.

Problem 5.10 Suppose you place 9 rings on the 3 mid fingers of your left hand (that is, not on your thumb or your pinky). How many different outcomes are possible if

(a) all the rings are identical, and no finger has more than 3 rings?

Answer

1.

Solution

There must be exactly 3 rings on each finger.

(b) all the rings are different, and no finger has more than 3 rings?

Answer

$9! = \frac{9!}{(3!)^3} \cdot (3!)^3 = 1680.$

Solution

Each finger has at exactly 3 rings, so we need to divide the rings into 3 arrangements of 3. However we also need to order the rings on each of the 3 fingers. Alternatively, note this is equivalent to just ordering all 9 rings.

(c) all the rings are identical, and no finger has more than 8 rings?

Answer

$\binom{9+3-1}{9} - 3 = 52.$

Solution

Use stars and bars to calculate the total number of outcomes and subtract the 3 ways where all the rings are on a single finger.

(d) all the rings are different, and all the rings are on a single finger?

Answer

$3 \cdot 9! = 1088640.$

Solution

There are 9! ways to order the rings, which can then be placed on one of the 3 fingers.

6 Solutions to Chapter 6 Examples

Problem 6.1 Prove the following algebraically:

(a) The symmetry formula.

Solution

This is routine.

(b) Pascal's Identity.

Solution

This is routine.

(c) The reduction formula.

Solution

This is routine.

Problem 6.2 Calculate the following. The identities may be useful in simplifying your answer.

(a) You have 10 friends and you give out (identical) copies of Monopoly to some of them. If no one gets more than one copy, how many ways are there to give out games to 5 or 6 of your friends?

Solution

$$\binom{10}{5}+\binom{10}{6}=\binom{11}{6}=462.$$

(b) There are 4 types of presents: (i) Monopoly, (ii) a basketball, (iii) a t-shirt, and (iv) chocolate cookies. (You have multiple identical copies of each of the 4 types.) You decide to create a gift collection (containing some or all of the presents above). How many ways are there to create the collection if it contains between 0 and 7 presents?

Answer

330

Solution

Each case is stars and bars, and then we can use the Hockey Stick Identity to help simplify the answer:

$$\begin{aligned}
& \binom{0+4-1}{0}+\binom{1+4-1}{1}+\cdots+\binom{7+4-1}{7} \\
= & \binom{3}{3}+\binom{4}{3}+\cdots+\binom{10}{3} \\
= & \binom{11}{4} \\
= & 330.
\end{aligned}$$

Problem 6.3 Prove the following using a combinatorial proof.

(a) The symmetry formula.

Solution

Picking which k objects you want is equivalent to picking which $n-k$ objects you do not want.

(b) The reduction formula. Hint: Consider a committee with a president.

Solution

Suppose you have n people and want to pick a committee of k people, and among these k people, one is chosen as president of the committee. If we pick the committee first and then the president from the committee, there are $\binom{n}{k}\cdot k$ ways. If we pick the president first and then the rest of the committee, there are $n\cdot\binom{n-1}{k-1}$ ways.

Problem 6.4 Answer each of the questions below in two different ways. Then expand the arguments to prove the remaining identities about binomial coefficients.

(a) Suppose you have one friend named Gwen, as well as 10 other friends. You want to invite a group of 5 friends out to dinner. How many different groups of friends could you invite?

Answer

$\binom{10}{4} + \binom{10}{5} = \binom{11}{5}$.

Solution

Consider cases based on whether Gwen is invited or not. The same argument works with n other friends and inviting a group of k to dinner to prove Pascal's Identity.

(b) Consider a city grid below:

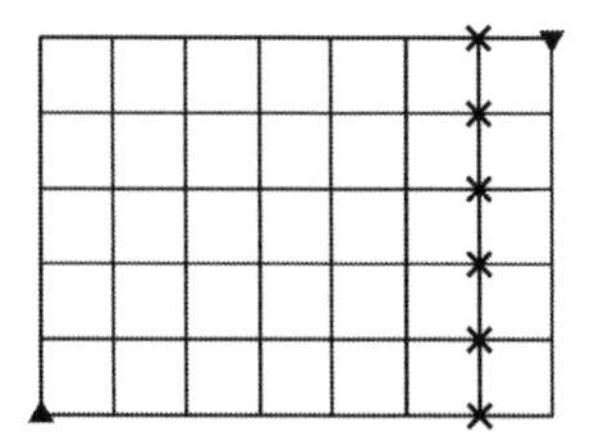

How many paths are there from the lower left triangle to the upper right triangle? Hint: The $\times$'s are in the diagram for a reason!

Answer

$\binom{6}{6} + \binom{7}{6} + \binom{8}{6} + \binom{9}{6} + \binom{10}{6} + \binom{11}{6} = \binom{12}{7} = 792.$

Solution

There are $\binom{12}{7}$ such paths. We can break these paths into cases based on which $\times$ they travel through *last* during the path. (If a specific $\times$ is the last traveled through, the path must have gone right from that $\times$, so there is only one way to complete the path to the upper right triangle.)

The same argument can be expanded to show the Hockey Stick Identity.

Problem 6.5 Use the following questions as a guide, give a proof of the binomial theorem.

(a) How many words can you form with $n-k$ $A's$ and k B's?

Answer

$\binom{n}{k}$.

Solution

The entire word is of length n, so choose where the B's go.

(b) Argue that, when expanding $(a+b)^n$, you get the sum of all words of length n consisting of a's and b's. For example, $(a+b)^2 = aa+ab+ba+bb = a^2+2ab+b^2$.

Solution

Note this combined with part (a) will provide a proof of the binomial theorem.

Problem 6.6 Simplify the following

(a) $\binom{n}{0}+\binom{n}{1}+\binom{n}{2}+\cdots+\binom{n}{n}$.

Answer

2^n.

Solution

Let $a=b=1$ and use the binomial theorem.

(b) $\binom{n}{0}-\binom{n}{1}+\binom{n}{2}+\cdots+(-1)^n\binom{n}{n}$.

Answer

0.

Solution

Let $a=1, b=-1$ and use the binomial theorem.

Problem 6.7 Simplify $\sum_{k=0}^{n} \frac{1}{k+1}\binom{n}{k}$.

Answer

$\dfrac{2^{n+1}-1}{n+1}$.

Solution

Use the reduction formula,
$\binom{n}{k} = \frac{k+1}{n+1}\binom{n+1}{k+1}$,
shift the index ($l = k+1$), and use the binomial theorem:

$$\sum_{k=0}^{n} \frac{1}{k+1}\binom{n}{k} = \sum_{k=0}^{n} \frac{1}{k+1}\frac{k+1}{n+1}\binom{n+1}{k+1} = \sum_{k=0}^{n} \frac{1}{n+1}\binom{n+1}{k+1} = \frac{1}{n+1}\sum_{k=0}^{n}\binom{n+1}{k+1}$$
$$= \frac{1}{n+1}\left[\sum_{l=0}^{n+1}\binom{n+1}{l} - 1\right] = \frac{2^{n+1}-1}{n+1}.$$

Problem 6.8 Find the coefficient of

(a) x^4y^5 in $(x+y)^9$.

Solution

$\binom{9}{5} = 126.$

(b) x^4 in $(x+3)^9$.

Answer

$\binom{9}{5} \cdot 3^5 = 30618.$

Solution

Let $y = 3$ and use the Binomial Theorem.

Problem 6.9 Find the constant term in the expansion of $\left(\sqrt{x} + \frac{1}{\sqrt{x}} - 2\right)^5$.

Answer

$(-1)^5\binom{10}{5} = -252.$

Solution

Let $y^2 = \sqrt{x}$, then the expression becomes

$$\left(y^2 + \frac{1}{y^2} - 2\right)^5 = \left(y - \frac{1}{y}\right)^{10},$$

so use the Binomial Theorem to find the coefficient of the constant term, where the constant term is $y^5 \cdot \left(\frac{1}{y}\right)^5 = 1$.

Problem 6.10 Prove that $\sum_{j=0}^{n} \binom{m}{j} \cdot \binom{n-m}{k-j} = \binom{n}{k}$. Hint: $(1+x)^n = (1+x)^m(1+x)^{n-m}$.

Solution

Following the hint, the right hand side is the coefficient of x^k in $(1+x)^n$. Then note the coefficient of x^k in $(1+x)^m(1+x)^{n-m}$ is the sum of the coefficients of x^j in $(1+x)^m$ times the coefficient of x^{k-j} in $(1+x)^{n-m}$.

7 Solutions to Chapter 7 Examples

Problem 7.1 Turn each recursive definition of a sequence given below into a general formula.

(a) $a_0 = -4, a_{n+1} = a_n + 2.$

Answer

$a_n = -4 + 2n.$

Solution

This is an arithmetic sequence.

(b) $a_0 = 8, a_{n+1} = a_n/2.$

Answer

$a_n = \frac{8}{2^n}.$

Solution

This is an geometric sequence.

(c) $a_0 = 1, a_{n+1} = (n+1) \cdot a_n.$

Answer

$n!.$

Solution

The first few terms are $1, 2, 6, 24, \ldots$.

(d) $a_0 = 0, a_{n+1} = a_n + n + 1.$

Answer

$\frac{n(n+1)}{2}.$

Solution

Note $a_n = 1+2+\cdot+n$ for $n \geq 1$.

Problem 7.2 Find recursive definitions of the following sequences. Writing out a few small examples might help! You do not need to prove your answers.

(a) Let F_n (for $n \geq 1$) denote the number of ways to write $n-1$ as the sum of 1's and 2's. For example, $3 = 1+1+1 = 2+1 = 1+2$ so $F_4 = 3$.

Answer

$F_{n+1} = F_n + F_{n-1}$.

Solution

The first few terms are $1, 1, 2, 3, 5, 8$.

(b) Let a_n (for $n \geq 1$) denote the number of ways to invite a group of your friends (assume there are n total friends) if you invite at least one friend but not all the friends.

Answer

$a_1 = 0, a_{n+1} = 2a_n + 2$.

Solution

The first few terms are $0, 2, 6, 14$.

Problem 7.3 Let $A = \{1,2,3,4,\ldots,10\}$ and $B = \{1,2,3,4\}$. Answer each of the following. Explain how you have already answered this type of question in the past.

(a) How many total functions are there from A to B? from B to A?

Answer

1048576, 10000

Solution

For each input, we must choose one of the outputs. Hence from A to B there are 4^{10} functions while from B to A there are 10^4 total functions.

(b) How many bijections are there from A to B?

Answer

0

Solution

A and B have different sizes, so there are no bijections between them.

(c) How many injections are there from B to A?

Answer

5040

Solution

For each member in B we need a different output in A, so this is a permutation. Hence there are

$$10 \cdot 9 \cdot 8 \cdot 7 = \frac{10!}{6!} = 5040$$

injective functions.

(d) How many surjections are there from A to B?

Answer

818520

Solution

There are 4^{10} total functions. Let events A, B, C, D denote functions with $1, 2, 3, 4$ respectively missing from the range. Using PIE we have

$$n(A \cup B \cup C \cup D) = \binom{4}{1} 3^{10} - \binom{4}{2} 2^{10} + \binom{4}{3} 1^{10}$$

so there are

$$4^{10} - \binom{4}{1} 3^{10} + \binom{4}{2} 2^{10} - \binom{4}{3} 1^{10} = 818520$$

surjective functions.

Problem 7.4 Let $A = \{1,2,3,4,5\}$.

(a) Give an example of a bijection between subsets of A of size two and subsets of A of size 3.

Solution

Let B be a subset of A of size 2. Map B to $A \setminus B$. For example, $\{1,4\} \mapsto \{2,3,5\}$. It is not hard to see this is a bijection.

(b) How many such bijections are there in part (a)?

Answer

10!.

Solution

Note that each set has size

$$10 = \binom{5}{2} = \binom{5}{3}.$$

If we fix the subsets of size 2, any rearrangement and pairing of the subsets of size 3 gives a bijection, so there are 10! bijections.

(c) Find a bijection between subsets of A of size 2 and two numbers chosen (repetition allowed) from $\{1,2,3,4\}$.

Solution

Suppose the subset of A is $\{a,b\}$ with $a < b$. Map this to the two numbers $a, b-1$. Convince yourself that this is a bijection.

Problem 7.5 Complete the following proofs combinatorially.

(a) Let $a_n =$ the number of subsets of $\{1,2,\ldots,n\}$. (So $a_n = 2^n$.) Prove combinatorially that $a_n = 2 \cdot a_{n-1}$.

Solution

Given a subset of $\{1,2,\ldots,n-1\}$, we create a subset of $\{1,2,\ldots,n\}$ by either adding n or not adding n (2 choices).

(b) Let $a_n =$ the number of permutations of $\{1,2,\ldots,n\}$. (So $a_n = n!$.) Prove combinatorially that $a_n = n \cdot a_{n-1}$. Hint: If you permute $\{1,2,\ldots,n-1\}$ how many "spaces" are available for n?

Solution

Following the hint, given a permutation of $\{1,2,\ldots,n-1\}$, there are n spaces created by the $n-1$ numbers. Placing n in any of the spots yields a permutation of $\{1,2,\ldots,n\}$.

Problem 7.6 Carol and Tom are getting married. They are going through their friends and deciding if (i) the friend is invited to the wedding and the reception, (ii) the friend not invited to the wedding but invited to the reception, or (iii) not invited at all. Assume they invite at least one of their friends to the reception and let a_n denote the number of ways they can give out invitations to n friends.

(a) Give a general formula for a_n.

Answer

$a_n = 3^n - 1$.

Solution

Note there are 3 choices for each friends, so in total 3^n ways. However, there is 1 way no one gets invited, so we have $a_n = 3^n - 1$.

(b) Give a recursive formula for a_n. Prove your answer combinatorially.

Answer

$a_1 = 2, a_{n+1} = 3a_n + 2$.

Solution

Given a group of invitations among n friends (so at least one is invited) there are 3 ways for the $n+1$st friend. This gives a total of $3 \cdot a_n$ ways. With the new friend however, there are also two possibilities where the new friend is the only one invited to the wedding or reception. Hence we have

$$a_{n+1} = 3a_n + 2$$

as needed.

Problem 7.7 Three identical standard dice are thrown. How many possible rolls are there? (That is, we only care about how many of each number are rolled.)

(a) Give a solution using stars and bars.

Answer

$\binom{3+6-1}{3} = \binom{8}{3} = 56.$

Solution

Let x_1 be the number of times the face "1" appears, x_2 be the number of times the face "2" appears, and so on. Thus $x_1+x_2+x_3+x_4+x_5+x_6=3$, where x_i are non-negative numbers, so use stars and bars.

(b) Give a solution using a bijection. Hint: Consider your answer to part (a).

Answer

$\binom{8}{3} = 56.$

Solution

For each possible outcome, there is exactly one way to put the three numbers in increasing (actually non-decreasing) order. For example, $(1,2,4)$, $(3,3,5)$, or $(2,2,2)$. Let the three numbers be a,b,c and assume $1 \le a \le b \le c \le 6$. In order to count easily, we use the following 1-1 correspondence (called *bijection*): let $A=a, B=b+1, C=c+2$. This way we have $1 \le A < B < C \le 8$. The number of choices for (A,B,C,D) is $\binom{8}{3}$. Since each choice of (A,B,C) corresponds to exactly one choice of (a,b,c), and vice versa, the answer is $\binom{8}{3} = 56$.

Problem 7.8 In Problem 7.2 we saw the *Fibonacci* sequence, with recursive definition $F_n = F_{n-1} + F_{n-2}$, and $F_1 = F_2 = 1$ (or sometime $F_0 = 0$). We also saw that $F_n =$ the number of ways to write $n-1$ as a sum of 1's and 2's. Use this characterization to prove the recurrence $F_n = F_{n-1} + F_{n-2}$.

Solution

Note any sum of 1's and 2's either starts with a 1 or a 2. Therefore, removing the first term of a sequence summing to $n-1$ gives either a sequence summing to $n-2$ or $n-3$.

Problem 7.9 Suppose n friends go to a party. They each wear a coat. However, as they are leaving, they each randomly grab a coat. Let $a_n =$ the number of ways the friends can leave so that *none* of them have their own coat.

(a) $a_0 = 1$ by convention. Find a_1, a_2.

Solution

$a_1 = 0, a_2 = 1$.

(b) Prove combinatorially that $a_n = (n-1)\cdot(a_{n-1} + a_{n-2})$. Hint: With n people, the first person can choose one of $n-1$ hats. Then consider two cases.

Solution

The first person has $n-1$ hats (not their own) to choose from. Suppose he takes the kth hat (where $k = 2, \ldots, n$). Consider cases based on whether k takes the first person's hat (leaving us with $n-2$ people) or not (and pretend that the first person's hat is now k's, so we have $n-1$ people).

Problem 7.10 Let F_n denote the nth Fibonacci number. Prove

$$F_1^2 + F_2^2 + \cdots + F_n^2 = F_n F_{n+1}$$

geometrically. Interpret F_k^2 as the area of a square with side length F_k and $F_n \cdot F_{n+1}$ as the area of an F_n by F_{n+1} rectangle.

Solution

Following the idea above, start with a $F_n \times F_{n+1}$ rectangle. Removing a $F_n \times F_n$ rectangle, we are left with a $F_n \times (F_{n+1} - F_n) = F_n \times F_{n-1}$ rectangle. Continuing in this manner, we remove squares of size $F_{n-1}, F_{n-2}, \ldots$ until we remove the entire rectangle. The $n = 5$ diagram is below ($F_1 = 1, F_2 = 1, F_3 = 2, F_4 = 3, F_5 = 5, F_6 = 8$):

5^2 3^2 1^2 1^2 2^2

8 Solutions to Chapter 8 Examples

Problem 8.1 Suppose you roll 2 fair six-sided dice. Let A be the event that the first die is a 6, B the event that the sum of the two rolls is 7, and C be the event that the sum is even.

(a) Calculate $P(A)$, $P(B)$, $P(C)$.

Solution

$P(A) = \frac{6}{36} = \frac{1}{6}, P(B) = \frac{6}{36} = \frac{1}{6}, P(C) = \frac{18}{36} = \frac{1}{2}$. Note the sample space has size $6 \cdot 6 = 36$. Note for C, the sum is even if both rolls are even or both rolls are odd.

(b) Find $P(A \cap B)$.

Answer

$P(A \cap B) = \frac{1}{36}$.

Solution

Note that $A \cap B$ means the first is a 6 and the second is a 1.

(c) Find $P(B \cup C^c)$.

Answer

$\frac{18}{36} = \frac{1}{2}$.

Solution

Note that C^c means the sum is odd. Since 7 is odd, $B \cup C^c = C^c$.

(d) Find $P(A \cap C)$.

Answer

$\frac{3}{36} = \frac{1}{12}$.

Solution

Note that $A \cap C$ means the first is a 6 and thus the second must be even. Hence $n(A \cap C) = 3$.

Problem 8.2 Suppose you flip a fair coin 8 times.

(a) What is the probability you get 5 heads?

Solution

$\binom{8}{5}/2^8 = \frac{56}{256} = \frac{7}{32}$.

(b) What is the probability you get an equal number of heads and tails?

Solution

$\binom{8}{4}/2^8 = \frac{70}{256} = \frac{35}{128}$.

(c) What is the probability you get more heads than tails?

Answer

$\frac{93}{256}$.

Solution

Note there are $256 - 70 = 186$ outcomes with an unequal amount of heads and tails. Half of these (93) have more heads.

Problem 8.3 Suppose you randomly pick a number on the number line between -3 and 5.

(a) What is the probability the number is positive?

Answer

$\frac{5}{8}$.

Solution

The total length is $5-(-3)=8$, and the length of what we want is $5-0=5$.

(b) What is the probability that the number squared is greater than 4?

Answer

$\frac{4}{8}=\frac{1}{2}$.

Solution

Note the number squared is greater than 4 if the number is greater than 2 (length $5-2=3$) or less than -2 (length $-2-(-3)=1$). This has a total length of $3+1=4$.

Problem 8.4 Suppose Jack and Jill both randomly come to school between 9 AM and 1 PM.

(a) What is the probability that Jack comes to school before 10 AM?

Answer

$\frac{1}{4}$.

Solution

Consider the outcomes on a number line. There are 4 total hours, and 1 of those hours is before 10 AM, so the probability is $\frac{1}{4}$.

(b) What is the probability Jack and Jill both come to school before 11 AM?

Answer

$\frac{1}{4}$.

Solution

Consider a square in the xy-plane, where the x-axis corresponds to when Jack arrives and the y-axis corresponds to when Jill arrives. This gives the picture below

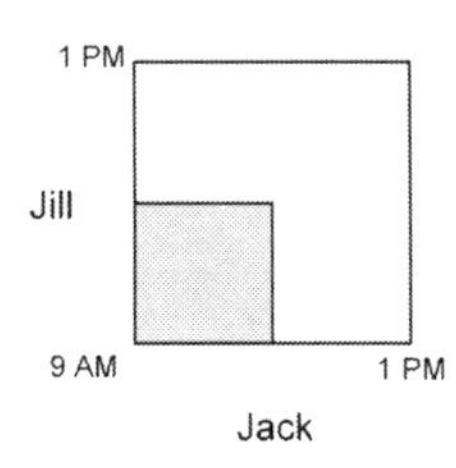

where the shaded region is Jack and Jill both coming to school before 11 AM. This region is $1/4$th of the entire square.

(c) What is the probability Jack comes to school before Jill?

Answer

$\frac{1}{2}$.

Solution

Use a method similar to part (b). We have the picture

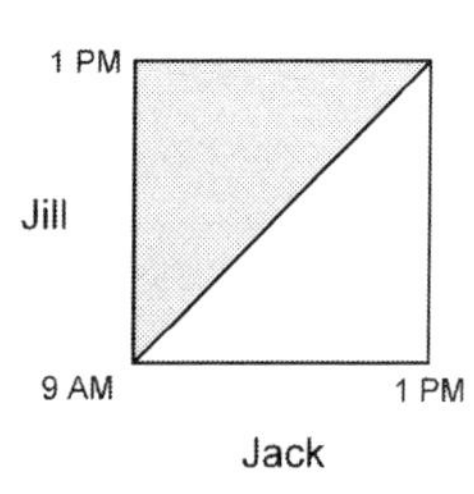

where now the shaded region is Jack coming to school before Jill. This region is $1/2$ of the entire square.

(d) What is the probability Jack and Jill come to school within 1 hour of each other?

Answer

$\frac{7}{16}$.

Solution

Use a method similar to part (b). We have the picture

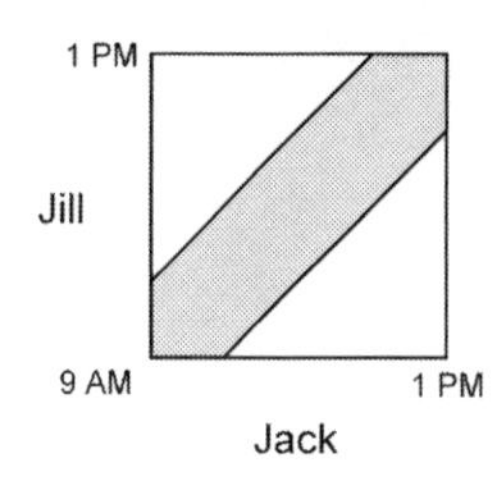

where now the shaded region is Jack and Jill arriving within 1 hour of each other. This region is $7/16$ of the entire square.

Problem 8.5 Suppose I pick a number from $\Omega = \{1,2,3,4,5,6,7,8\}$. However, each number is not equally likely to be chosen: the probability I pick 1, 3, or 5 is .15, the probability I pick 2 or 4 is .05, the probability I pick 6 is .25, and the probability I pick 7 is the same as the probability I pick 8.

(a) Find $P(7)$ and $P(8)$.

Solution

All the probabilities have to add up to one, hence $P(7)+P(8) = 1-3\cdot.15-2\cdot.05-.25 = .2$. Thus, $P(7) = P(8) = .1$.

(b) Find the probability I pick an even number.

Answer

.45.

Solution

We simply add up the probabilities of all the even numbers.

(c) Find the probability I pick a prime number.

Answer

.45.

Solution

The primes are $2,3,5,7$.

Problem 8.6 Suppose $A, B \subseteq \Omega$ such that $P(A) = .4, P(B) = .4$ and $P(A \cup B^c) = .8$. Find $P(A^c \cap B^c)$. Hint: Draw a Venn Diagram!

Answer

.4.

Solution

Note that $(A \cup B^c)^c = B \cap A^c$, so $P(B \cap A^c) = .2$. Hence, $P(A \cap B) = .2, P(A \cap B^c) = .2$. Thus, $P(A^c \cap B^c) = 1 - .2 - .2 - .2 = .4$.

Problem 8.7 Suppose you are dealt a five card hand from a standard deck of 52 cards (with 13 ranks $2, 3, \cdots, 10, J, Q, K, A$ and 4 suits: hearts, diamonds, clubs, and spades). Find the probability of:

(a) a full house (3 cards of one rank, 2 cards of another).

Answer

$\binom{13}{1}\binom{12}{1}\binom{4}{3}\binom{4}{2}/\binom{52}{5}$.

Solution

We first choose the rank for the three of a kind, then the rank of the pair. Then we choose the 3 suits for the three of a kind, then the 2 suits for the pair.

(b) exactly two pair. (That is, not four of a kind or a full house.)

Answer

$\binom{13}{2}\binom{11}{1}\binom{4}{2}^2\binom{4}{1}/\binom{52}{5}$.

Solution

First pick the ranks of the pairs, then the final rank. Then choose the suits.

(c) a flush (all cards of the same suit).

Answer

$\binom{4}{1} \cdot \binom{13}{5} / \binom{52}{5}$.

Solution

We choose the suit and then 5 of the 13 cards from that suit.

(d) a straight (all 5 ranks in a row, so $A, 2, 3, 4, 5$ up to $10, J, Q, K, A$).

Answer

$10 \cdot 4^5 / \binom{52}{5}$.

Solution

Note the ranks of the cards in a straight are determined by the lowest card (which can be $A, 2, 3, \ldots, 10$, so 10 choices). Then each rank can be any of the 4 suits.

Problem 8.8 Suppose you have line segments of length $1, a, b$, where a, b are real numbers chosen randomly between 0 and 2. What is the probability you can form a triangle using the three line segments.

Answer

$\frac{5}{8}$.

Solution

Note using the triangle inequality, we must have $a+b > 1, b+1 > a, a+1 > b$. Graphing where all 3 of these inequalities is true yields the following (restricted to $0 \leq a, b \leq 2$)

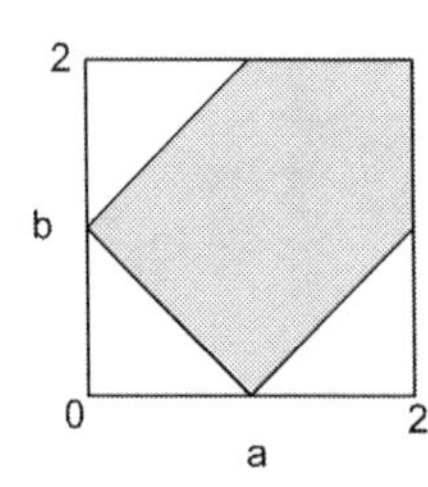

The entire square has area 4, and the shaded region has area $\frac{5}{2}$.

Problem 8.9 A bag contains 6 red, 5 green, and 4 yellow balls. You pick 4 balls at once (so in no particular order).

(a) What is the probability you get 3 red, 1 green, and 1 yellow ball?

Answer

$\binom{6}{3}\binom{5}{1}\binom{4}{1}/\binom{15}{4}$.

Solution

We pick which red, green, and yellow balls are chosen.

(b) What is the probability you get 0 yellow balls?

Answer

$\binom{11}{4}/\binom{15}{4}$.

Solution

We have 11 non-yellow balls, so we choose 5 of them.

(c) What is the probability you get all 4 balls of the same color?

Answer

$\left[\binom{6}{4}+\binom{5}{4}+\binom{4}{4}\right]/\binom{15}{4}$.

Solution

Consider the three cases: all red, all green, all yellow.

(d) What is the probability you do not get all red and green balls?

Answer

$1-\left[\binom{11}{4}/\binom{15}{4}\right]$.

Solution

Use the complement. Note this is the opposite to part (b).

Problem 8.10 Suppose Alice, Bob, and Charlie plan to meet for dinner. They each randomly show up to dinner between 5:00 and 6:00 PM. Alice will wait for Bob and Charlie to show up. Bob will wait for Charlie but not Alice. Charlie will not wait. What is the probability they have dinner together?

Answer

$$\frac{1/6}{1} = 1/6.$$

Solution

Let a, b, c stand for the time Alice, Bob, and Charlie arrive for dinner, measured as a fraction of an hour after 5 PM. They will have dinner if $a < b < c$. We can then view their arrival times as a coordinate (a, b, c). The inequality $a < b < c$ is a triangular prism with base area $1/2$ and height 1 (so volume $1/6$). As their arrival times form a cube of volume 1, the probability is $1/6$.

9 Solutions to Chapter 9 Examples

Problem 9.1 There is an urn with 5 green, 6 red, and 4 yellow balls. You pick 3 balls without replacement (that is, without putting the balls back after each pick). Hint: Does this remind you of dealing cards from a deck?

(a) Let A be the event you have at least 3 green balls, and B be the event you have at least 2 green balls. Find $P(A|B)$.

Answer

$$\frac{\binom{5}{3}}{\binom{5}{2}\binom{10}{1}+\binom{5}{3}}.$$

Solution

Note the denominators of $P(A\cap B)$ and $P(B)$ are the same, so they cancel. For $P(B)$ we consider two cases (2 or 3 green balls).

(b) Let A be as above, and B be the event that the first 2 balls picked are green. Find $P(A|B)$.

Answer

$$\frac{3}{13}.$$

Solution

If we know the first three balls are green, just consider picking one ball from the remaining 3 green and 10 not green balls.

(c) Compare your answers. How would they compare to $P(A)$? (You do not need to calculate $P(A)$.)

Solution

The answers are different (note in (b) we know *which* two balls are green). Both answers are definitely larger than $P(A)$.

Problem 9.2 Jack is planning to have dinner with Miss Muffet, and will do so as long as he arrives before the spider (who scares her away). Suppose Jack arrives randomly between 5 and 6 PM, and the spider arrives randomly between 5 and 7 PM.

(a) What is the probability Jack has dinner with Miss Muffet?

Answer

$\frac{3}{4}$.

Solution

View the total outcomes as 1×2 rectangle.

(b) Given that Jack arrives before 5:30, what is the probability he has dinner with Miss Muffet?

Answer

$\frac{7}{8}$.

Solution

We now restrict ourselves to the left half of the rectangle.

(c) Given that Jack has dinner with Miss Muffet, what is the probability he arrives before 5:30?

Answer

$\frac{7}{12}$.

Solution

We now restrict ourselves to the upper $3/4$ of the rectangle.

Problem 9.3 Suppose an urn contains 2 red, 3 green, and 4 yellow balls. You pick one ball, note it, and replace it (back in the urn). You do this a total of 6 times, what is the probability:

(a) You get exactly 3 green balls.

Answer

$\binom{6}{3}\left(\frac{3}{9}\right)^3\left(\frac{6}{9}\right)^3$.

Solution

On any given pick we have a $3/9$ change of getting a green ball (so $6/9$ chance of not getting a green ball). We also need to decide which of the picks yield a green ball.

(b) You get all 6 balls of the same color.

Answer

$$\left(\frac{2}{9}\right)^6+\left(\frac{3}{9}\right)^6+\left(\frac{4}{9}\right)^6.$$

Solution

Consider cases based on which color. Note since we are replacing the balls this is possible!

(c) You get the same number of each color.

Answer

$$\frac{6!}{(2!)^3}\left(\frac{2}{9}\right)^2\left(\frac{3}{9}\right)^2\left(\frac{4}{9}\right)^2.$$

Solution

Think of the possibilities as a word consisting of R,R,G,G,Y,Y.

(d) You get exactly 3 green and 2 yellow balls.

Answer

$$\frac{6!}{3!\cdot 2!}\left(\frac{3}{9}\right)^3\left(\frac{4}{9}\right)^2\left(\frac{2}{9}\right).$$

Solution

Note this implies we also have 1 red ball. We then proceed similarly to part (c).

Problem 9.4 Suppose boys and girls are equally likely and each child's gender is independent from the others. Let A be the event that the family has at least one child of each gender. Let B be the event the family has at most one girl. Are A and B independent if:

(a) the family has two children.

Answer

Not independent.

Solution

$P(A) \cdot P(B) = \frac{2}{4} \cdot \frac{3}{4} \neq P(A \cap B) = \frac{2}{4}.$

(b) the family has three children.

Answer

Independent.

Solution

$P(A) \cdot P(B) = \frac{6}{8} \cdot \frac{4}{8} = P(A \cap B) = \frac{3}{8}.$

Problem 9.5 Suppose you are dealt two cards from a standard deck. Find and compare the following probabilities:

(a) the probability the first card is a club.

Answer

$\frac{13}{52}$

(b) the probability the second card is a club.

Answer

$\frac{13}{52} \cdot \frac{12}{51} + \frac{39}{52} \cdot \frac{13}{51} = \frac{13}{52}.$

Solution

Note the first card is either a club or not a club, so use the law of total probability.

(c) the probability the second card is a club, given that the first card is a club.

Answer

$\frac{12}{51}$.

Solution

Note you've already answered this in part (b)!

(d) the probability the first card is a club, given that the second card is a club.

Answer

$\frac{(13/52)(12/51)}{(13/52)} = \frac{12}{51}$.

Solution

Note we've already calculated the probability the second card is a club in part (b).

Problem 9.6 Suppose you flip an unfair coin. If you get heads, you roll a fair die twice. If you get tails, you roll a fair die three times.

(a) If you get tails, what is the probability the sum of the rolls is 8?

Answer

$\frac{21}{216}$.

Solution

Using stars and bars, there are $\binom{7}{2} = 21$ positive solutions to $a+b+c=8$. Note all of these correspond to dice rolls, so there are 21 ways for the sum to be 8.

(b) Suppose the probability of tails is 4 times that of the probability of heads. Find the probability the sum of all the rolls is 8.

Answer

$\frac{114}{1080} = \frac{19}{180}$.

Solution

First note $P(H) = 1/5$. Let S be the event that the sum is 8. Then $P(S) = P(H) \cdot P(S|H) + P(T) \cdot P(S|T)$. This equals $\frac{1}{5} \cdot \frac{5}{36} + \frac{4}{5} \cdot \frac{21}{216}$.

Problem 9.7 Suppose that at a given time, 1 in 500 people have strep throat. Further, suppose we have a test for strep throat so that: (i) If the person has strep throat the test works 98% of the time, (ii) If the person does not have strep throat, the test works 96% of the time. Suppose we randomly take someone and test them for strep throat. Let S be the event that a person has strep and $\oplus$ be the event a person tests positive. Thus, S^c is a person not having strep throat and $\oplus^c = \ominus$ is a person testing negative.

(a) Find $P(S), P(\oplus|S), P(\ominus|S)$. Hint: This is a reading comprehension question as well as a math one!

Answer

$P(S) = 1/500 = .002$, $P(\oplus|S) = .98$, $P(\ominus|S) = 1 - P(\oplus|S) = .02$.

Solution

Note that $P(\ominus|S)$ is the "rate of false negatives."

(b) Write out the Bayes' Theorem formula for $P(S|\oplus)$ and calculate this probability.

Answer

Note, first, that $P(S^c) = .998$. Now, to say that the test works 96% of the time if the person does not have strep throat means that $P(\ominus|S^c) = .96$. Hence the rate of false positives is $P(\oplus|S^c) = 1 - P(\ominus|S^c) = .04$. With this information, Bayes' Theorem gives us

Solution

$$\begin{aligned} P(S|\oplus) &= \frac{P(S) \cdot P(\oplus|S)}{P(S) \cdot P(\oplus|S) + P(S^c) \cdot P(\oplus|S^c)} \\ &= \frac{.002 \times .98}{.002 \times .98 + .998 \times .04} \\ &\approx 0.047 \end{aligned}$$

This says the "accuracy rate" of the test, $P(S|\oplus)$, is about 4.7%.

Problem 9.8 Suppose you roll a fair six-sided die. You then flip a coin the number of times shown on the die.

(a) What is the probability you get exactly 5 heads and 0 tails?

Answer

$\frac{1}{192}$

Solution

Note this means you must have rolled a 5. The probability of rolling a 5 is $\frac{1}{6}$, and then rolling 5 heads in a row has probability $\left(\frac{1}{2}\right)^5 = \frac{1}{32}$. Therefore the answer is

$$\frac{1}{6} \cdot \frac{1}{32} = \frac{1}{192}.$$

(b) What is the probability you get exactly 5 heads?

Answer

$\frac{1}{48}$

Solution

Note that you must have rolled either a 5 or a 6.

If you rolled a 5 and then got 5 heads, the probability is $\frac{1}{192}$ by Part (a).

If you rolled a 6 and then got 5 heads and 1 tails, the probability is

$$\frac{1}{6} \cdot \binom{6}{5} \left(\frac{1}{2}\right)^6 = \frac{1}{64}.$$

So the answer is

$$\frac{1}{192} + \frac{1}{64} = \frac{1}{48}.$$

(c) Given that you get exactly 5 heads, what is the probability you rolled a 5?

Answer

$\frac{1}{4}$

Solution

Based on Part (a) and Part (b), the probability of rolling a 5 and then 5 heads is $\frac{1}{192}$, and the probability of rolling a 6 and then 5 heads and 1 tails is $\frac{1}{64}$. Given that 5 heads are obtained, then the probability of rolling a 5 is found by the Bayes' Formula:

$$\frac{\frac{1}{192}}{\frac{1}{192}+\frac{1}{64}} = \frac{1}{4}.$$

Problem 9.9 Suppose Jack arrives randomly between 5 and 6 PM, the spider arrives randomly between 5 and 7 PM, and Jack gets to have dinner with Miss Muffet as long as he shows up before the spider. Let A be the event that Jack has dinner with Miss Muffet and B be the event that Jack and the spider show up within an hour of each other. Are A and B independent?

Answer

Not independent.

Solution

We have $P(A) \cdot P(B) = \frac{3}{4} \cdot \frac{3}{4}$ and $P(A \cap B) = \frac{1}{2}$ and they are not equal.

Problem 9.10 Suppose Billy is nervous for his first airplane flight. So nervous that he cannot remember his assigned seat when he is the first to board the plane. He randomly sits in one of the 100 seats. Further, every other passenger is very polite, and if Billy (or anyone else) is in their seat will simply sit randomly in one of the remaining.

Suppose you are the last to board the plane. What is the probability you get your assigned seat? Hint: First examine the problem for smaller numbers of seats and try to find a pattern/relationship between them!

Answer

$\frac{1}{2}$.

Solution

Let A_n be the event you get your seat if there are n seats on the plane. Thus the question is asking for $P(A_{100})$. $P(A_2) = 1/2$, as there is a $1/2$ chance Billy takes your seat. For A_3, consider three cases: (i) Billy takes his own seat (you automatically get your own seat), (ii) Billy takes your seat (no chance you get your seat), or (iii) Billy takes the other persons seat. In case (iii) note we can then view the other person as the new Billy and use A_2. Thus,

$$P(A_3) = \frac{1}{3} \cdot 1 + \frac{1}{3} \cdot 0 + \frac{1}{3} \cdot P(A_2) = \frac{1}{3} + \frac{1}{6} = \frac{1}{2}.$$

Note the same idea works for A_4, as if Billy takes one of the other person's seats, we are either reduced to A_2 or A_3 which we already know both have probability $1/2$, so

$$P(A_4) = \frac{1}{4} \cdot 1 + \frac{1}{4} \cdot 0 + \frac{2}{4} \cdot \frac{1}{2}.$$

This pattern continues, so in general we have

$$P(A_n) = \frac{1}{n} \cdot 1 + \frac{1}{n} \cdot 0 + \frac{n-2}{n} \cdot \frac{1}{2} = \frac{1}{2}.$$

Made in United States
Troutdale, OR
09/26/2023